RF-31849

SPACE AGE

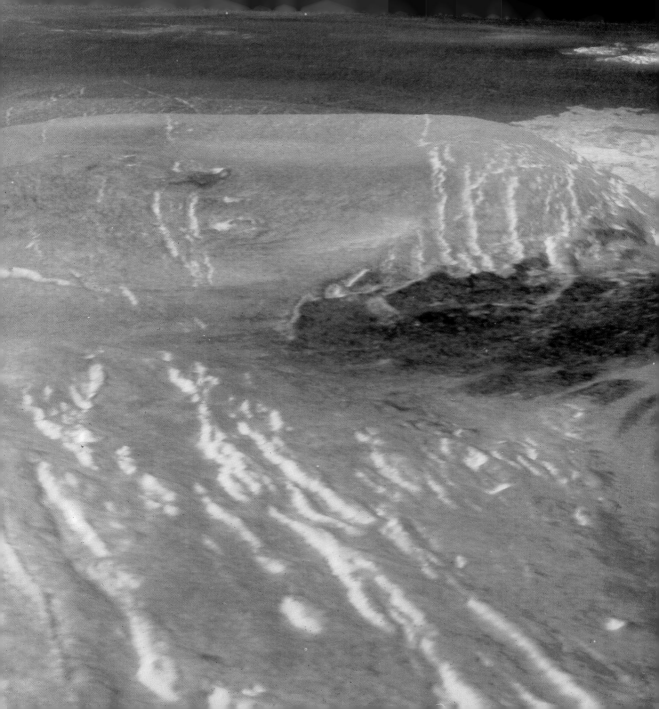

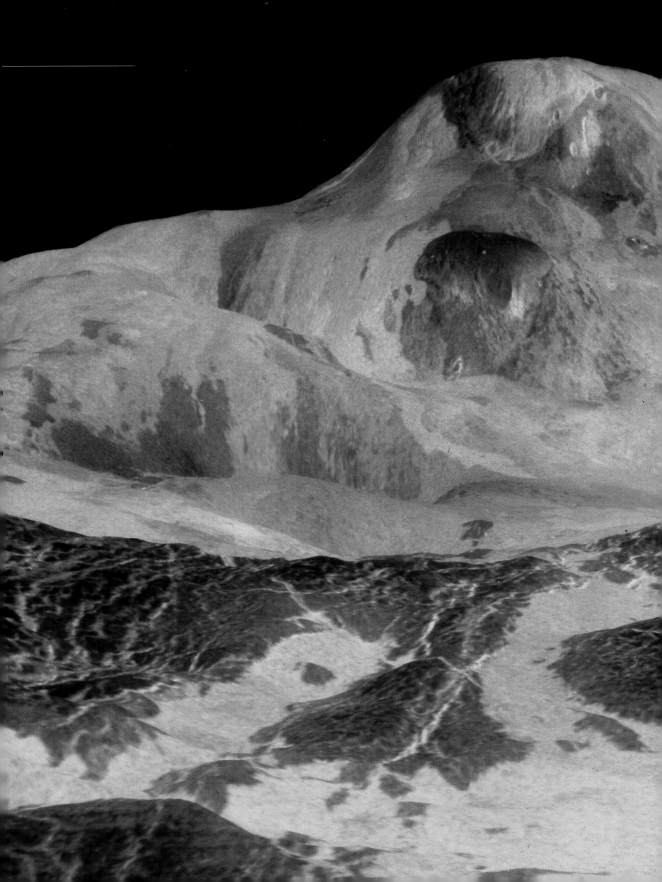

SPACE AGE

William J. Walter

Random House   New York

Grateful acknowledgment is made to Warner/Chappell Music for permission to reprint an excerpt from "Let's Do It" by Cole Porter. Copyright 1928 By Warner Bros. Inc. Copyright renewed. All rights reserved. Reprinted by permission.

Library of Congress Cataloging-in-Publication Data

Walter, William J.
    Space age / by William J. Walter.—1st ed.
        p.   cm.
    Companion vol. to Space age, a WQED/Pittsburgh, NHK/Japan TV series produced in association with the National Academy of Sciences and NASA.
    Includes index.
    ISBN 0-679-40295-0
    1. Outer space—Exploration.   2. Outer space—Exploration—United States.   3. Artificial satellites.   4. Mars (Planet)—Exploration.   5. Space industrialization.   I. Space age (Television program)   II. Title.
TL790.W35   1992   629.4—dc20   92-6041

Manufactured in the United States of America
9   8   7   6   5   4   3   2
First Edition

Contents

Foreword   ix

Acknowledgments   xiii

1   Dreams to Reality   5

2   The Explorers   51

3   Quest for Planet Mars   115

4   Mission to Planet Earth   175

5   Celestial Sentinels   213

6   New Frontiers   255

EPILOGUE   The Shape of Things to Come   303

Bibliography   313

Credits   324

Index   327

For Mary,
with all my love

Foreword

T he *Space Age* book and television series had its beginnings four years ago in discussions at station WQED in Pittsburgh that had been prompted by scientist Herb Friedman. Herb's pioneering efforts helped invent a new field called space physics, and eventually led to the discovery of an "invisible universe" filled with a stunning array of truly violent objects, many in the throes of creation and destruction. As a result, the ordered view of the universe we had clung to since Copernicus was finally shattered in this century.

We were in good hands with this pioneer. He told us of a growing effort on the part of the world's scientific organizations to dub 1992 the International Space Year, in honor of the quincentennial of the first voyage of Columbus, and the thirty-fifth anniversary of the International Geophysical Year of 1957–58, an unprecedented global exploration of the Earth that also saw the launch of *Sputnik*. In fact, *Sputnik* would overshadow much of what the I.G.Y. achieved. Ironically, before the tumultuous global change of 1989, the I.S.Y. was also to commemorate the seventy-fifth anniversary of the Russian Revolution! How could we know? Space was becoming more international in focus and the I.S.Y. would be an effort to help promote public awareness of the growing importance of space in all our lives. The late Senator Spark Matsunaga of Hawaii, who helped invent the idea for the I.S.Y., believed that it would generate broad-based, sustained interest in space activities of a sort never yet attained. While the I.S.Y. has not yet achieved Senator Matsunaga's dream of an international exploratory mission to Mars, it has already spawned a number of popular communication efforts, like this one, that may inspire Senator Matsunaga's dream in this and succeeding generations.

At the time, WQED/Pittsburgh had just finished a successful, Emmy-winning collaboration with the National Academy of Sciences for the *Planet Earth* series and its companion book. Together we immediately set to work, with some generous development support from NASA, to invent a series that would provide a new look at space: not just a rehash of the space race and exploration stuff, but a new perspective on how influential many aspects of the Space Age have been on the shape of our lives and the world. Few efforts have yet tried to capture the meaning of the Space Age, and this would be an opportunity to communicate this fresh perspective to millions of people.

William Walter, "Chip" as we all know him, joined us early to do research, help articulate the editorial approach to the series, and write the first television treatments. We were joined by a distinguished group of advisors from the National Academy of Sciences, headed by Edward Stone of the Jet Propulsion Laboratory, chief scientist of the spectacularly successful Voyager spacecraft missions, to help fine tune our ideas. When the series was funded, former astronaut Sally Ride, now with the University of California, San Diego, chaired a larger group that appointed experts to advise us on each of our themes. Their generosity and counsel have helped provide a veracity and credibility to our efforts with the television series. We were very happy, when the time came to create a companion book for *Space Age*, that Chip eagerly accepted our offer to be the writer. We are grateful for Chip's skill in bringing alive both history and character, and in explaining the extraordinary consequences of the Space Age, which range from the practical to the cosmic.

History seems to tell us that the real revolutions wrought by exploration are usually not in the discovery, but in its consequences—the ideas and new perspectives that we bring back are what ultimately change us. Consider some of the most important changes from the age of space:

• Our first views of Earth from space gave us a profound sense that the planet is a precious living system, and began a dramatic reassessment of our relationship with it.

• Space science was invented, and stunning surprises have resulted from the many robot tours of the planets of our solar system.

• The surprising finding that there is more than meets the

human eye when it comes to the universe. By getting above our atmosphere and looking beyond the visible spectrum, we discovered an unbelievably violent universe that has forever changed our view of the cosmos.

• The creation of sophisticated rocket ships began the manned exploration of the heavens. We have now begun to understand the challenges we must face to probe our frontiers.

• The building of global satellite communications networks made it nearly impossible to keep one part of the world from another. Human visions were expanded; the global village became a reality. Unexpectedly, the global order that powered much of the Space Age was overwhelmed by these devices and the messages they carried.

• And maybe for the first time in human history the tools of war actually did keep a tenuous peace. We are just beginning to understand the decisive role that "orbital power," embodied in surveillance, weather, resource, and other satellites, has already played in the shaping of our world.

These are just some of the big changes. But the Space Age is also a great human story—one of the great adventure stories of history—full of intrigue, global rivalries, secrecy, breakthroughs, surprises, heroes and heroines and heroics, investigatory brilliance, bravado, huge risk, and profound failure, and an increasing understanding of who we are and where we fit in the universe.

Most of all, what emerges from this book is an unstinting sense of optimism, a celebration of whatever drives us into space, this "seemingly senseless curiosity within," which may be the savior of the human race. These efforts are a tribute to something that is really the "right stuff" of humankind.

We are indebted to many people. At WQED, Executive Vice President Tom Skinner, who eagerly supported and guided the project through its many phases; WQED president Lloyd Kaiser; Ted Bogosian, our inventive series producer who led a very talented staff in the creation of the television programs; producers Joe Seamans, Gary Hines, Peter Argentine, Akira Yoshizawa, Jim Golway, Doug Bolin, John Rubin, Gail Willumsen, and their respective teams; special effects supervisor John Allison; stock photography coordinators Michael Mushlitz and Jessica Thompson; Mark Friedman, who managed WQED's manuscript and photography preparation; and Jeanne Paynter, Peggy Zapple, Charlene Haislip, and Peggy Rafalko of WQED, who were with this all the way. We are grateful to Hidemi Hyuga of NHK, Japan, for his support and friendship throughout the production, and to Sosuke Yasuma who began the production collaboration between WQED and NHK. WQED's literary agent Victoria Pryor kept us on track. At the National Academy of Sciences, Barbara Valentino again expertly coordinated the many scientists and advisors, all of whom made such an important contribution to these efforts. Frank Owens at NASA helped launch our project, yet never once asked anything of us editorially. NASA, the National Science Foundation, and the Corporation for Public Broadcasting are supporting our formal educational materials, which will assure usage by American schoolchildren so in need of enhancements in scientific learning and literacy. Funding for our television series comes from viewers who contribute to the stations of the Public Broadcasting Service and from the Corporation for Public Broadcasting, with special thanks to Jennifer Lawson and Don Marbury and to Barry Chase, formerly of PBS; the National Science Foundation and Hyman Field; the Arthur Vining Davis Foundations; the U.S. Department of Energy; and corporate funder McDonnell Douglas. We are grateful to our other international coproduction collaborators, SVT-1 in Sweden, TROS of the Netherlands, and ABC in Australia. The growing internationalization of television production for major series reflects many of the themes found in *Space Age.*

Arthur C. Clarke, one of the true visionaries of the Space Age (he predicted a global girding of communication satellites as early as 1945) testified on videotape before the U.S. Congress in 1984, suggesting that it would be "absurdly optimistic to hope that by Columbus Day 1992, the United States and the Soviet Union will have emerged from their long winter of sterile confrontation." If anything is true of the Space Age it is that things change unexpectedly. One thing we can count on, however, is also echoed by Clarke, "Where there is no vision, the people perish." He continued by saying, "Men need the mystery and romance of new horizons almost as badly as they need food and shelter. In the difficult years ahead, we should remember that the Snows of Olympus lie silent beneath the stars, waiting for our grandchildren." In some fashion, hopefully, our efforts will help to continue this marvelous journey.

Gregory Andorfer
Executive Producer, Space Age
Vice President, National Programming, WQED/Pittsburgh

Acknowledgments

Writing a book is both a humbling and an exhilarating experience. It reminds you how little you know, but gives you the opportunity to learn so much. This book, the companion volume to the six-part Public Broadcasting System series produced by WQED/Pittsburgh, has been no exception. For me the project began one early spring day in 1988 in Los Angeles, California, when Greg Andorfer, WQED's vice president for national programming, asked if I would be interested in developing a series on the exploration of space. While I felt that my expertise in this subject was astoundingly shallow, I was willing to dive in. We began with a few simple questions: Why had the human race come to conceive and build machines that explore beyond our own planet? How had this amazing series of endeavors we call the Space Age changed us and where might they lead? As we got under way the story proved to be even more fascinating than we had suspected, and when the series was funded, I felt very lucky to have the chance to write the book.

I am grateful to Greg, not only for the invitation to become involved in *Space Age,* but for his help and insight, genuine love of science, and refreshing sense of wonderment, none of which flagged throughout the project.

An undertaking that involves nothing less than the entire universe requires the talent and hard work of many people, and this book would never have been completed without the aid of many kind and talented people around the country. Lloyd Kaiser, president of WQED, and Tom Skinner, executive vice president, deserve thanks for their support and confidence, especially during the final weeks of editing the manuscript. Their understanding and patience as tight deadlines approached is deeply appreciated.

While I was writing in Los Angeles, the hardworking staff for the series was in Pittsburgh doing the grueling work and research necessary to put a topflight documentary series on the air. I want to thank Ted Bogosian, the series producer, for his thoughts over the past eighteen months. Busy as they were, show producers Jim Golway, Peter Argentine, and Joe Seamans of WQED, and Akira Yoshizawa of NHK, all managed to find time to provide transcripts, copies of video interviews, and stacks of research material for me, aid which made my job much easier. Others on the series staff took time from their unrelenting schedules to track down articles, supply leads, and answer questions as they worked to shape and plan their programs. Harry Gural, Doug Bolin, Ando Yuichiro, and Carey Strelecki deserve special thanks. Transmitting information, or even knowing where it could be found, was often crucial, and when it absolutely, positively had to be delivered, Dana McBurney, the series production assistant, made sure that it was. Her patience and good nature never flagged even when deadlines loomed.

Beautiful photography and artwork grace this book, and all of those images were found by Michael Mushlitz, photo editor, and his assistant Jessica Thompson who worked ungodly hours for months tracking down the very best photos and paintings for *Space Age,* many of them rarely or never published before. Michael, forever patient, intelligent, and uncompromising, always kept in mind the editorial message of each passage when searching for an image, and Jessica's effervescence and thoroughness remained constant. I deeply appreciated the steady counsel, judgment, and good humor of Mark Friedman, WQED's design director, especially during the last hectic weeks as we worked to choose the final photos and wrestled with design ideas. He and his computers even helped with the delivery of the final manuscript. John Allison and his award-winning special effects unit also deserve thanks; some of their magical images are in this book.

I also want to thank Jeanne Paynter, the series project manager and marketing director, for her unremitting efforts to make potential funders see the light, but even more for her wit and energy and confidence. Jeanne never seems daunted by anything, no matter how overwhelming. Peggy Rafalko, WQED's national programming assistant, was involved from the very beginning, and was always helpful, insightful, and supportive on every level, from logistical to personal. When a problem needed to be solved, she solved it. Charlene Haislip, the series unit manager, was also always kind and helpful, and never seemed too tired or busy to take on whatever arcane problem I brought to her. In New York I would also like to thank my agent Peter Sawyer and WQED's agent Victoria Pryor for their encouragement and faith in me.

This project is the result of an association with the National Academy of Sciences and would never have become a reality without the help of this amazing organization. Barbara Valentino, director of the Academy's television office, worked from beginning to end to assemble and coordinate meetings with top scientists, historians, and journalists from around the world for the Academy panels that advised on the series itself. Many of these people, beginning four years ago when research on this project first began, set aside many hours to talk with me. It was a great joy to be in the company of such sparkling minds and stimulating thinking. For their help, guidance, and insight throughout this project, I would particularly like to thank Herb Friedman, Jim Head, Lynn Margulis, Frank McDonald, Robert McCormick Adams, Michael McElroy, Edward Stone, John Noble Wilford, Gene Shoemaker, and Guyford Stever, all people at the very top of their fields. This isn't the first time many of these people have given me the benefit of their knowledge, and I hope it isn't the last. For their comments on the manuscript, I would also like to thank Robert Parks of the Jet Propulsion Laboratory, now a retired aerospace consultant, satellite pioneer John McLucas, Richard Somerville of the Scripps Institution of Oceanography, John Niehoff of SAIC Corporation, and Kevin Burke of the National Research Council, and Joseph Allen, president of Space Industries International.

Space Age could never have been completed on time without the help of Phyllis Kaelin, my research assistant. Her special talents enabled her to find and assemble and organize thousands of pages of material on subjects ranging from microbes to nuclear propulsion. Phyllis plowed through stacks of books, articles, and papers, braved Los Angeles traffic, and faced down recalcitrant librarians in a never-ending battle to supply the raw material for this book and track down answers to questions of cosmic import. Never daunted, she was always cheerful and positive and hacked her way through a jungle of information with joy, thoroughness, and intelligence. I have thanked her many times for her help, but it's a particular pleasure to do it here.

I also want to thank my parents for their encouragement,

which began long before this project ever did, and express my deepest gratitude to Mary, my wife, and to my daughter, Molly. Molly has passed three quarters of her first two years with a father incessantly tapping a computer keyboard. Now when asked what her dad does for a living, she replies, "He pushes buttons." Mary put up with my absence, my babbling about rocketry, planetary science, and physics, yet always remained encouraging and patient, willingly reading rough drafts and inevitably providing advice and insight that was precisely on target. No matter how humbled, exhilarated, or "spacey" I would become, she, at least, always remained down to Earth.

Chip Walter
Pittsburgh/Los Angeles, 1992

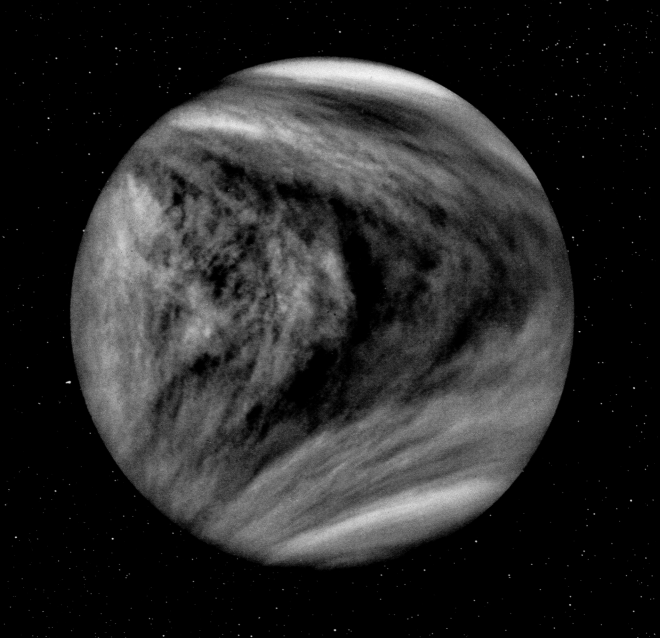

"What is now proved true, was once only imagined."

—William Blake

Dreams to Reality

Top: Konstantin Tsiolkovsky at the age of thirty-seven. He was toying with the idea of rockets, but hadn't yet conceived of them.

Kaluga at the turn of the century; a small, simple town outside of Moscow.

At the turn of the century the world was riding the crest of an unprecedented wave of technological innovation. Thomas Edison had already invented the electric light, the phonograph, and the movie projector; the Wright brothers were building their first "flying machine" in their Dayton, Ohio, bicycle shop; and in Italy Guglielmo Marconi was putting the finishing touches on a "wireless" that would soon send messages clear across the Atlantic. All of these advances would deeply reshape the lives of people everywhere, but none would change the world more than the Space Age, an era that was also in the process of being born in a small Russian town called Kaluga, about one hundred miles southwest of Moscow. It was here in 1898 that a deaf, forty-one-year-old arithmetic teacher named Konstantin Tsiolkovsky calculated precisely how human beings might depart Earth to explore other worlds.

That Tsiolkovsky had worked out his calculations entirely on his own with only a vague awareness of the innovations occurring around him made his feat all the more amazing. Partly because of his deafness, he had kept to himself throughout his life, rushing home each night after teaching at the local school and huddling in his small workshop where he developed theories on extraterrestrial life and the power of the sun, and where he ultimately conceived of a device he came to call a "reaction machine," a liquid-fuel rocket ship that could be used to propel humans beyond Earth. These sperm-shaped cylinders, he imagined, would

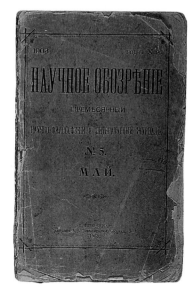

The cover of the Russian Journal *Nauchnoyie Obozrenie (Science Survey)* 1903, in which Tsiolkovsky's landmark paper "The Investigation of World Spaces with Reaction Machines" was published. "I have worked out various aspects of the problem of ascending into space with the aid of a reaction machine, rather like a rocket," he wrote.

someday seed the universe and utterly transform the human race.

Five years before Orville Wright managed to skim a scant 120 feet along the sands of Kitty Hawk, Tsiolkovsky wrote:

> I have just worked out various aspects of the problem of ascending into space with the aid of a reaction machine, rather like a rocket.
>
> The scientifically ... verified mathematical conclusions indicate the feasibility of an ascent into space with the aid of such machines and, perhaps, the establishment of settlements beyond the confines of the Earth's atmosphere.

Five years passed before Tsiolkovsky found anyone willing to print a piece of thinking this outlandish, but finally, in 1903, the editor of *Nauchnoyie Obozrenie (Science Survey)* agreed to publish his paper under the simple title, "The Investigation of World Spaces with Reaction Machines."

At a time when people had never seen rockets used for anything other than fireworks, Tsiolkovsky's bizarre prophecies were so astonishing that they could only be described as madness. Yet a mere fifty-eight years later, Yuri Gagarin, one of Tsiolkovsky's countrymen, rode the nose of one of these "reaction machines" into orbit and was able to radio back something no one had ever been able to say before: "I can observe the Earth ... one can see everything."

Of course many others had contemplated space travel before. Three hundred years before Christ, Antonius Diogenes wrote of an imaginary trip to the moon, and in the seventeenth century Dutch mathematician Christiaan Huygens wrote that the universe brimmed with other worlds. Authors from Daniel Defoe to Cyrano de Bergerac had written stories about journeys beyond Earth, and as a boy Tsiolkovsky himself had been inspired by two Jules Verne novels that described a human launch into space and a trip around the moon.[1]

Years later as a university student in Moscow, Tsiolkovsky had convinced himself that he had actually found a way to accomplish what Verne had only fantasized about—a journey beyond

---

[1]   The books, written in the 1860s, were *From the Earth to the Moon* and its sequel *Round the Moon,* both of which described journeys in a "cylindro-projectile" shot from a monstrous cannon in Florida not far from where Cape Canaveral is now located. They were immensely popular and set the stage for Verne's other masterpieces, *Journey to the Center of the Earth* and *Twenty Thousand Leagues Under the Sea.*

Earth. It would be made by a great centrifugal wheel that would spin a ship into space. He walked the streets for hours one night trying to puzzle out the details, but by morning he realized that his centrifugally powered ship would never work. Nevertheless, he kept wrestling with the problem, landscaping the theoretical terrain, until finally, years later, he arrived at an altogether different method for space travel: the rocket—a conclusion that seems obvious today, but was far from apparent at the time.

Tsiolkovsky was an amazing man, not simply because he conceived these unearthly ideas alone and out of touch with the rest of the world, but because he blended his wild-minded passion for space with the most hardheaded calculation, and produced not simply an invention but a vision. It wasn't the rocket itself that was so important to him, it was what the rocket could do. He believed that someday more advanced versions of his reaction machine would scatter the human race across the Milky Way, and make humans the inhabiters of countless worlds. In the process of this great migration, he predicted, we would build self-sustaining space stations, explore the moon and Mars, and design earth-orbiting solar power stations, space suits, and artificial gravity machines. He imagined a time when humans would re-engineer the environments of other planets, populate the asteroid belt beyond Mars, and communicate with extraterrestrial beings—all ideas that have become centerpieces of modern space exploration.

Nobody knew of a space age in Tsiolkovsky's day, not a soul had seen it coming, but his insights meant that somewhere out on the cultural and scientific mantle, the landscape was shifting. Impossible dreams were about to be fulfilled, and for the first time a primordial fascination with the heavens would soon be satisfied. In a letter written to a friend in 1911, Tsiolkovsky predicted, "Man will not always stay on earth. The pursuit of light and space will lead him to penetrate the bounds of the atmosphere, timidly at first, but in the end to conquer the entire solar system." Fittingly, it was Tsiolkovsky himself who took the first real step.

There isn't a corner of the planet, not a continental nook or cranny, that humans don't inhabit or haven't explored. Other animals migrate—the swallows return annually to San Juan Capistrano, great herds of wildebeests thunder across the Serengeti each year, and every spring the vultures come to roost in Hinckley, Ohio—but human wanderings are rarely so predictable. We are

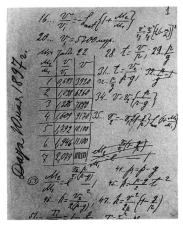

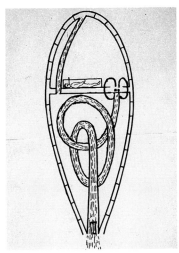

Top: The formula that launched a thousand ships. Tsiolkovsky calculated precisely the power needed to launch a rocket into space. The formula determined the rocket's speed at any moment, the speed of gas outflow, the mass of the rocket, and the amount of fuel used.

Bottom: A sperm-shaped rocket that would seed the solar system. From "The Investigation of World Spaces with Reaction Machines," Tsiolkovsky's first drawing of a reaction machine. A human reclines in the upper chamber.

Illustrations for Jules Verne's science fiction adventures, *From the Earth to the Moon* and its sequel *Round the Moon*. Verne's travelers were blasted to the lunar surface from an enormous cannon and found the sights fascinating. Despite certain scientific shortcomings, his fiction inspired rocket pioneers like Tsiolkovsky and Oberth.

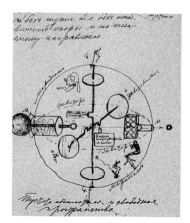

Above: A preliminary sketch of a ship Tsiolkovsky drew for a diary entry written in 1883, entitled "Free Space."

Right: Victorian comfort on the way to the moon. A library, a dog, and an open observation window in an illustration from Verne's *From the Earth to the Moon.* Weightlessness was not a problem and space suits were not required.

Opposite: Looking back on Earth on the way to the moon. One hundred years later that view would revolutionize our attitudes about our own world.

Человѣчество не останется вѣчно на землѣ, но, въ погонѣ за свѣтомъ и пространствомъ, сначала робко проникнетъ за предѣлы атмосферы, а затѣмъ завоюетъ себѣ все около-солнечное пространство

К. Ціолковскій

ramblers, in both mind and body, and this characteristic has some-how transformed us from a ragged band of primate-hunters into a race that probes other planets.

This need to explore is so strong that we've fabricated re-markable tools to extend our reach. We invented the wheel, built the great sailing ships of the seventeenth century, created the Iron Horse, the airplane, and the rocket, a machine designed to take us beyond the world that brought us into being. These tools are linked at the deepest level with our ability to dream up new possibilities.

Some anthropologists believe a direct connection exists be-tween the complex toolmaking talents that have ensured human survival, and the idea-making abilities that separate our species from other creatures. There are obvious parallels between the combining of different objects to make tools, and the blending of information and experience to create new insight. Dreams and tools have a common origin. Two million years ago, when *homo habilis,* the first hunter and toolmaker, lived along the lake shores of east Africa whacking flint axes out of rock, he had already become much more than a wild animal; he was living proof of the connection between manipulating matter and processing knowl-edge. His predatory skills had made a prey of knowledge, though the answers rarely provided a single calorie of nourishment in return. This implacable curiosity, something scientist and author Jacob Bronowski has called "a rage for knowledge," is the hall-mark of the human race. It has taken us to the moon, enabled us to dispatch probes that paw at the secrets of the planets, and brought us closer to an understanding of our place in the uni-verse.

Our earliest attempts to explain the universe, however, rarely reflected the facts: Our ancestors summoned up all manner of gods, giants, and unlikely beings out of the bestiary of their imag-inations to explain the unknown. The Egyptians believed the Earth was an egg guarded at night by "a great white bird," the moon. The ancient Peruvians imagined a boxlike world with a ridge-shaped roof where the Great God lived. Some civilizations thought that the stars were the souls of the dead, and that the Milky Way was a path made bright with their souls. The Bushmen of the Kalahari Desert still call this ribbon of light the backbone of the night, as if we lived within the belly of a great beast and could look up to see its spine high above. To the Greeks the sun

Top: Tsiolkovsky with his grandchildren. By the end of his life his writings had anticipated space stations, multistage rockets, space suits, and artificial gravity.

Bottom left: The Clovis were the first inhabitants of the American continent twelve thousand years ago. They made stone points out of rock just as *Homo habilis* had.

Bottom right: "Man will not always remain on Earth. The pursuit of light and space will lead him to penetrate the bounds of the atmosphere, timidly at first, but in the end to conquer the entire solar system." From a letter written by K. E. Tsiolkovsky to his friend B. N. Vorob'yev on August 12, 1911.

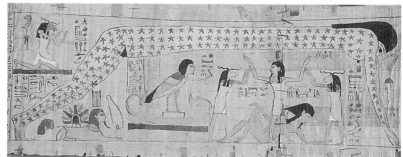

Top right: Zoroaster and the Magi. Hermes Trismegistus hands the *Corpus Hermeticum,* a legendary book of magic, to the learned men of the East and West. The Ionians borrowed from the astrology of the Magi to develop mathematics, which helped pave the way to the Space Age.

Bottom right: Nut, the Egyptian Goddess of the sky. For the Egyptians, and many other ancient cultures, myths explained the cosmos, but they made better stories than accurate descriptions.

was a fiery chariot driven by Apollo, and to the ancient Hindus the top half of the Earth sat on the backs of four elephants who, in turn, stood on the shell of an enormous tortoise that swam in the seas of the world.

Myths cloak the skeletons of the human psyche and symbolize our deepest fears and highest hopes; without them we would not be human. But long ago, they represented the primal *desire* for knowledge, not knowledge itself. In the end, myth made entertaining explanations, but poor proof.

Among the first people to recognize this difference between proof and explanation were the Ionians, a society of consummate wanderers who lived in a group of settlements that hugged the shores of Asia Minor six centuries before Christ. Ionians were traders, prodigious travelers who had little time for ritual and myth, but valued knowledge deeply, and expropriated it shamelessly from other cultures along their routes of trade. They were the first to use the stars as navigational beacons, for example, a trick they based on information from the astrological charts of Zoroastrian priests. They also produced the first navigational maps in the West, and sailed their ships as far north as the Black Sea and as far west as the unknown and dangerous waters of the Atlantic. The Ionians believed in keeping a close watch on the world, and endeavored to support their observations with solid proof, a tradition that led them to conclude that the cosmos operated according to certain immutable laws, not the whims of fickle gods.

This refusal to accept elaborate stories in place of hard fact laid the groundwork for the scientific method that emerged a thousand years after the Ionian culture disappeared. Renaissance thinkers like Copernicus, Galileo, and Newton built on their methods and ushered in the sciences of astronomy and physics, which became guideposts for exploring the universe. The insights of these men, in turn, created the foundations of the Space Age, and Tsiolkovsky drew heavily upon them. His rocket, however, was a concept that fused dream making and toolmaking in a new and spectacular way and represented a leap in human progress that no one foresaw. Today, less than one hundred years after the publication of Tsiolkovsky's paper on reaction machines, twenty-two nations have active space programs. Tens of billions of dollars, rubles, yen, marks, and francs are pouring out of national coffers into rockets, satellites, and interplanetary robots. It is a stated goal of the United States to install human outposts on the moon and

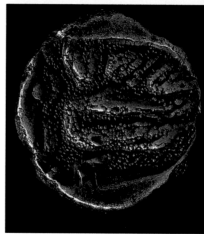

Two sides of an Ionian coin map. Used as currency by Ionian traders, this coin also provided the traveling trader with a map of an area in Asia Minor near Ephesus.

Above: Sir Isaac Newton. Newton's laws of motion explain why the moon orbits the Earth. He observed that for every action there is an opposite and equal reaction—the foundation of rocket flight.

Top right: The Ptolemic model of the universe. Ptolemy believed that the Earth was at the center of the universe and that the sun and other planets revolved around it. He saw the Earth as a solid, unmoving body, and created this model to explain why the sun and stars appeared to travel across the sky.

Bottom right: The Copernican model of the universe. In 1543 Copernicus challenged the traditional view of the universe established by Ptolemy in the second century. He proposed that the Earth was not at the center of the solar system, but that like the other planets, Earth rotated about the sun.

Mars, and not long ago the now-defunct Soviet Union also had similar projects afoot. However, the old Soviet space program has fallen upon uncertain times, and creative and daring work in human and robotic exploration being done in Russia today faces a difficult future.

But just as the old Soviet programs have become unstable, those of the Japanese and European space agencies have grown more vigorous. Europe is designing a laboratory for microgravity experiments that will orbit Earth, and various Japanese conglomerates have developed plans for Martian colonies and resorts on the moon. New discoveries as well as new plans are being made. Several probes launched by the United States—bright little robotic emissaries from Earth—are on their way to, or already exploring, other worlds, sending back information across millions of miles of space. Six thousand satellites orbit Earth, tracking changes in its weather, seas, and land, shuttling vast amounts of information around the globe, even informing us against our own destruction. And the first two in NASA's series of four Great Observatories are now stationed in orbit bringing diverse aspects of the universe in for close inspection. The cosmic background explorer (COBE), another space observatory, may already have resolved one of the central questions in science — how did the universe come into being. It recently revealed tantalizing evidence of the big bang.

THE SOVIET SPACE PROGRAM: BON VOYAGE?

It may come as a shock to the great pioneers of Soviet space exploration, but the once mighty Soviet space program no longer officially exists. The programs that it set in motion still constitute some of the most advanced technological undertakings in the world, involving tens of thousands of technicians, some of the world's most brilliant scientists and engineers, scores of research centers and launch sites, and of course Mir, the Earth's only working space station, but the future is uncertain.

The old space program now falls under the uncertain sway of the Commonwealth of Independent States (CIS), made up of eleven (when this book went to press) participating republics that were once part of the Soviet Union. The Commonwealth agreed on January 13, 1992, to coordinate the management and funding of space activities through a plan that is similar to the arrangement devised by the European Space Agency (ESA). ESA's member nations vote on which projects to fund, and then each government is invited to invest in those projects according to how important they feel they are. They then share in the contracts and benefits of the program according to their original investment.

In the new Commonwealth, however, agreements are complicated by the chaotic military situation. Unlike the United States, which has both a civilian and a military space program, the Soviet program was always entirely controlled by the U.S.S.R.'s Ministry of Defense. The Commonwealth's new military is now organized under what is called Joint Strategic Armed Forces, but some nations, notably Ukraine, Azerbaijan, and Moldova favor independent armies. As a result it is unclear who exactly controls reconnaissance, surveillance, and communication satellites as well as uncounted other installations that are considered to be military. Nor has it been decided whether the individual members of the Commonwealth will have equal access to these resources.

Late in 1991 there was talk that some republics might begin charging others for the use of space facilities on their territory. Kazakhstan, for example, had considered charging Russia for use of the Baikonur Cosmodrome, where all manned flights are launched, but now the new agreement states that critical facilities such as the Cosmodrome will be availabe to all other signatory states.

The chaos surrounding the dissolution of the Soviet Union has also postponed plans to send U.S. astronauts to the Mir space station in July 1992 and similar agreements with other countries are also up in the air. A complete disintegration of the space program could represent a stupendous loss of income and state-of-the-art technology for the new Commonwealth. Before the Soviet Union dissolved there had been plans to establish a large-scale materials processing unit on Mir, and to conduct tests to manufacture crystals, pharmaceuticals, and alloys in weightlessness. But the operation of Mir, the jewel of the old space program, has been turned over to a quasi-governmental organization called the Energia Scientific Industrial Company, which reports to the

Russian Ministry of Industry, and although Energia guarantees
that Mir will operate through 1992, the space station's exact
fate remains a mystery.

How has the human race gotten itself into a situation where
the sum total of its affairs—its politics, science, art, myth, eco-
nomics, and invention—have conspired to enable us to depart the
planet of our origin? The story begins with Tsiolkovsky, but em-
braces a much larger cast of astonishing characters who have followed him. Among them are two men who emerged early in this
century equally as spellbound by the dream of spaceflight as he.
One was an American physics professor named Robert Hutchings
Goddard, the other a Transylvanian mathematics instructor named
Hermann Oberth. That three people, each from a different part of
the world, who were completely unknown to one another, independently conceived of interplanetary rockets and the dreams that
go with them, says something about the inevitability of space exploration. It is as though forces in science, technology, and culture
had reached a global, critical mass, and these men represented
the first evidence that we were destined to explore beyond our
own planet.

When these men began to formulate their theories at the
turn of the century, precious little geography remained unexplored, and new frontiers were in demand. The American West
was settled. The sources of the Nile had been discovered and the
African interior filleted by adventurers like Richard Burton, David
Livingstone, and Henry Stanley. The ends of the Earth themselves
were fair game as Perry, Amundsen, Byrd, and Scott prepared to
close in on the North and South Poles. The world, it seemed, was
shrinking at last under eons of unrelenting human inquisition.

All of this only seemed to confirm what Robert Goddard had
proclaimed in his high school graduation speech in 1904: ". . . it
has often proved true that the dream of yesterday is the hope of
today and the reality of tomorrow." Or as he would also often say
later, "If there isn't a law against it, it will happen."

Anyone familiar with the roots of rocketry knows the story of
Robert Hutchings Goddard, part straightlaced Yankee, part mad
wizard: a man of surpassing passion and engineering skill and an
eccentric whose great accomplishments were undone by his own
unequivocal desire for secrecy.

A picture of Goddard taken when he was a physics professor at Clark University in Worcester, Massachusetts, reveals these contradictions. He stands at a blackboard in his tweeds and boiled collar, chalk in hand, his bald, egg-shaped pate gleaming, and a hint of a smile playing under his toothbrush mustache. This is Goddard the New England Yankee—an austere, unassuming instructor imparting a nugget of wisdom. But a closer inspection reveals something amazing on the blackboard—a drawing of Earth with a sketched rocket arcing off into the slate-black void toward the chalk-drawn moon. Goddard thought mad and visionary thoughts, but kept them to himself. His conservative side deplored the attention that his rocketry work drew, so rather than publicize his ideas he hid his dreams of interplanetary travel just as an alchemist guards his philosopher's stone.

At the age of thirty-six, for example, he sat down at his desk and sketched out an idea he called "The Ultimate Migration," the premise of which is that the sun is burning out, and the inhabitants of Earth are facing extinction. To escape, they build a fleet of enormous arks (or colonize and then redirect entire asteroids), and to ensure that at least one group will succeed in finding a new home, each ship carries its passengers off toward a different cluster of stars where a solar system with a habitable world might exist. On board each ship, Earth's survivors carry with them all human knowledge, literature, and art, and all of the technology necessary to seed a new civilization. His notes suggest that the ships might travel the immense distance to other stars using something he called "intra-atomic energy" (a prescient nod to atomic power, the mysteries of which hadn't yet been deciphered). Failing that, hydrogen and oxygen fuel, the kind used in many rockets today, combined with solar power would send the arks winging off to new star systems.

Goddard knew these odysseys might take a million years or longer, but he assumed the inhabitants would be placed in a state of suspended animation by reducing the protoplasm of the human body to a "granular state," enabling it to withstand the frigid cold of interstellar space. Every ten thousand to one million years the pilots would have to be "reanimated" to make certain that the ships were still on course. It would, after all, be unfortunate to make a wrong turn at Alpha Centauri and send the future of the human race hurtling seven or eight light-years out of its way.

After completing this astounding scenario, Goddard folded

Top: Goddard at the blackboard. Part straightlaced Yankee, part mad wizard, Goddard's inquiries into rocket travel began in high school, when he submitted an article to *Popular Science Monthly* titled "The Navigation of Space."

Bottom: The envelope in which Goddard enclosed his fantastic piece, "The Ultimate Migration." It was sealed with the warning "To be opened only by an optimist" scrawled across it.

the document and wrote on it, "To be opened only by an optimist." He then sealed it in an envelope and deliberately mistitled it "Special Formulae for Silvering Mirrors" so that no one would read it. Nine years later Goddard jotted down a postscript on his notes: "If the above is not possible or desirable, granular protoplasm, suitably enclosed, might be sent out of the solar system; this protoplasm being of such a nature as to produce human beings eventually, by evolution."

Goddard, of course, did more than scribble and squirrel away astonishing ideas. He was a top-notch physicist and a first-class engineer, a rare combination in one man. Like Tsiolkovsky he dedicated his entire life to the study of rocketry, but unlike Tsiolkovsky he actually built and launched the world's first liquid-fuel rocket. Because he was so solitary, many of his efforts were little known, but Goddard was happy to work alone—he was so much in the grip of his dream that he really had no choice.

From his boyhood, Goddard was fascinated with the possibilities of technology and driven to achieve greatness. His youth was sickly and isolated and spent in the company of thick stacks of reading material. By the time he was sixteen he had fallen two years behind in school, and was utterly frustrated. Convinced that he had to accomplish something meaningful and extraordinary in his life, he felt incapable of finding the one mission upon which he could focus his energy. Then, while recuperating from a kidney ailment, he read a serialized version of H. G. Wells's *The War of the Worlds* in the *Boston Post,* a story, he later recalled, that "gripped my imagination tremendously." Years later when he was fifty, Goddard even wrote Wells a letter telling the great writer how important the book had been in his life, explaining that "The spell [of the book] was complete about a year afterward, and I decided that what might conservatively be called 'high altitude research,' was the most fascinating problem in existence . . ."

The event that Goddard wrote to Wells about took place in October 1899 while he was recovering from yet another long illness. One afternoon he decided to go out for some fresh air and trim some of the trees in his grandmother's cherry orchard. From the tool shed he hauled out a saw and hatchet, headed off for the meadow behind her house, and climbed up into a tree.

He sat there for a long time, perhaps with *The War of the Worlds* simmering in his subconscious, and then in his mind's eye he envisioned a device capable of carrying him from Earth to Mars, powered—like the ship Tsiolkovsky had imagined in Moscow—by centrifugal force. Two horizontal shafts whirled at each end of this imaginary machine with two great weights at the tips of them. The one above spun faster than the one below, and because of its greater speed, it hoisted the ship out of the orchard and off to the stars. Years later he wrote, "I was a different boy when I descended the tree from when I ascended."

Goddard had finally stumbled upon the goal that would give his life purpose. In fact, he felt so strongly about his experience in the orchard that he forever marked it as his "Anniversary Day." He even revisited the tree later in life and took pictures of it with the ladder leaning against it, as if to recapture the power of his dream.

Goddard graduated from high school in 1905 and quickly took a science degree at the Worcester Polytechnic Institute. Ideas continued to ricochet in his mind like errant gunfire, but none resolved the mystery of extraterrestrial flight. First he thought his ship might escape Earth by being hurled into space like a bucket spun off a rope. Then he thought that an enormous nonrecoiling cannon might work, and later he hit upon the idea of using solar power or an electric gun that could propel the contraption skyward with a steady stream of charged ions. (Such engines are being designed today for interplanetary flight, but even now they qualify as futuristic.) None of these concepts, however, seemed feasible at the time.

Goddard grew frustrated, sometimes depressed. "God pity a one-dream man!" he wrote in his diary. After Worcester Polytechnic he took graduate and doctoral degrees in quick succession from Clark University, writing papers on diffraction and conductivity and current displacement, but still struggling in his spare time with the question of space travel. Then, almost ten years to the day after he experienced his first vision of spaceflight, Goddard finally realized, just as Tsiolkovsky had in 1898, that the best machine for traveling beyond Earth was a rocket. He won his first patent for a liquid-fuel "rocket apparatus" five years later, on July 7, 1914, and by the following autumn he was building his first prototypes in the physics shop at Clark University, where he had landed a part-time position as a physics professor.

The early rockets were of every imaginable stripe, designs based on anything from Chinese fireworks to flares to simple artillery. He had no real test range, so early in the New England

APPLICATION FILED OCT. 1, 1913.

1,102,653. Patented July 7, 1914

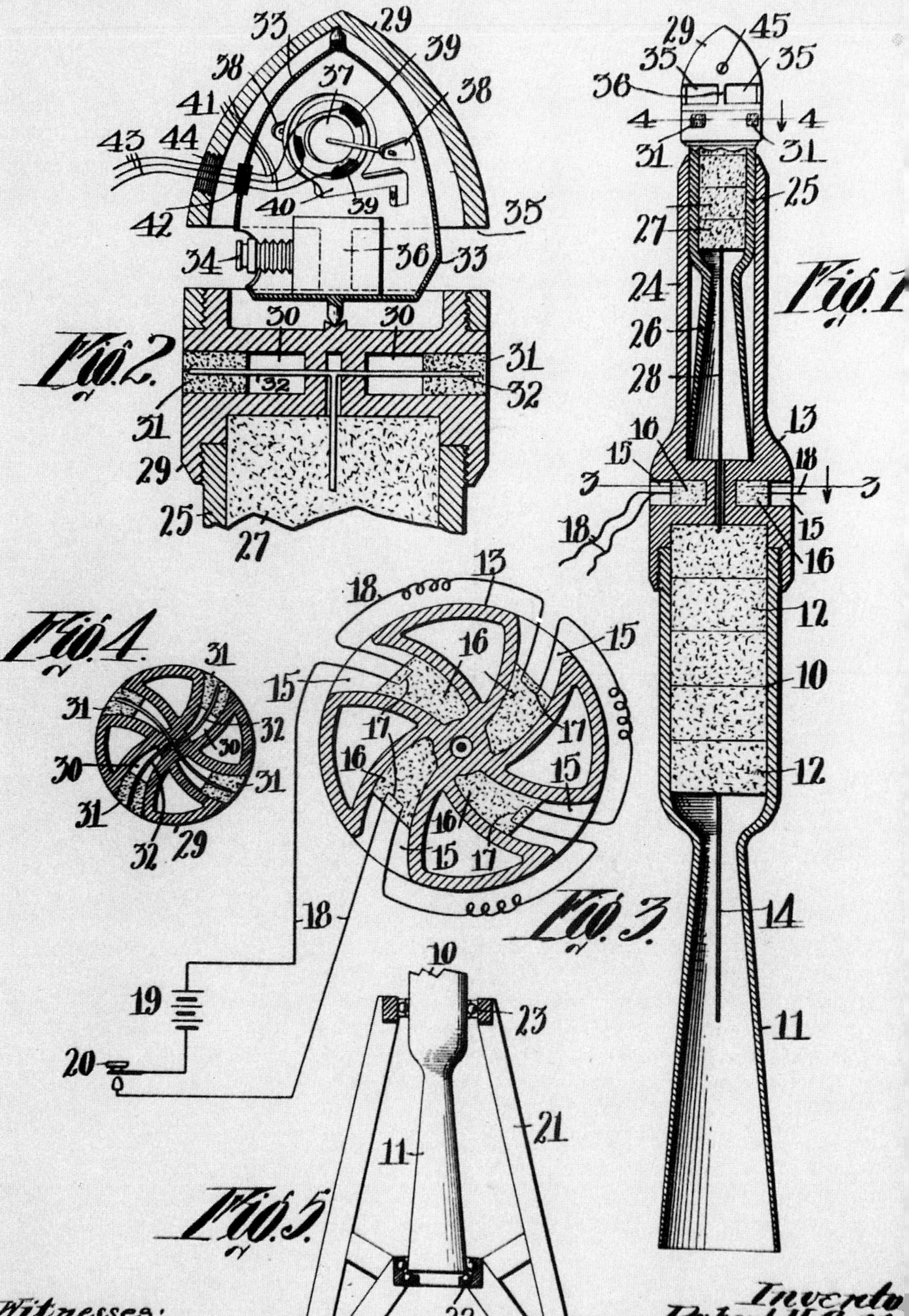
Fig. 1
Fig. 2
Fig. 3
Fig. 4
Fig. 5

Witnesses:

Inventor
Robert H. Goddard
by attorneys

mornings, before the sun rose, he would haul his inventions out to nearby Coes Pond and light up his prototypes and watch them arc out over the woods.

Subsequent rockets grew larger, as did the cost of building and testing them, and Goddard realized that if his research was to continue, he would need more money than he could ever supply from his own shallow pockets. So in September 1916 he approached the Smithsonian Institution for a research grant. The Smithsonian had generously funded the ground-breaking experiments that the great Samuel P. Langley had done at the turn of the century with early airplanes. Of course, Langley had been head of the Smithsonian at the time. Nevertheless, Goddard hoped that if he couched his proposal in scientific terms and was careful to keep his most outlandish ideas to himself, the Institution would show the same foresight with his rocketry work as it had shown in the past with Langley's aviation research. The letter he wrote therefore focused on technical details such as combustion efficiency, velocity of ejection, and the salutary results of his experiments, which he called "truly remarkable." His hope, he explained, was to invent a revolutionary way of exploring the upper atmosphere. Balloons were limited in their range, but a rocket with instruments packed in its nose could rise high above the planet and unveil atmospheric mysteries never before known.

Dr. Charles D. Walcott, the secretary of the Smithsonian, thought enough of the letter to write back asking Goddard to amplify his ideas. Goddard immediately sent him a detailed account of his theories in a volume bound in leather with marbleized end papers and an embossed gold title that read: "A Method of Reaching Extreme Altitudes."

The Smithsonian's experts liked the paper, and happened to have a source of funding on hand. A bequest made by Thomas G. Hodgkins of Long Island stipulated that half of his gift must be used to support atmospheric research—then and only then could the Smithsonian use the other half for whatever projects it wished. Within four months, Goddard received a check for the then-astronomical sum of $1,000, along with the even more astounding promise of $4,000 more "as he needed it."

While Goddard was making these strides in New England, Tsiolkovsky's efforts had stalled. Day after day, night after night, in his little laboratory on Georgiyevskaya Street, the aging Russian school teacher continued to hatch designs and plot the explora-

Goddard's 1914 patent for a multistage rocket.

tion of the solar system, even though hardly a soul knew of his existence, much less his work. From time to time a magazine or two published one of his pieces, but for the most part the manuscripts simply piled up in his small laboratory, amazing works of imagination that outlined descriptions of Earth from space, the effects of weightlessness and closed environmental systems, and a multistage rocket (a train really, that roared along a rising track toward the sky like a monstrous roller coaster and then shot into orbit dropping off one stage after another as it consumed its fuel).

On more than one occasion he had sent papers to the Russian Academy of Sciences, explaining his ideas and asking for support. No less than Dmitri Mendeleyev himself, the inventor of the periodic table of elements, had singled out Tsiolkovsky's work as superlative, but even then the Academy awarded him only 400 rubles, not the kind of money that launches space ages. However, the Russian government was never able to give Tsiolkovsky the attention he deserved. Then in 1917 Russia erupted in revolution: hordes of workers and soldiers revolted in Petrograd and forced the abdication of Czar Nicholas. A year later the Bolsheviks would be in power, Russia would pull out of the war, and the Czar's influence would be wiped out. It was only after the revolution, under the new Soviet government, that Tsiolkovsky, now in his early sixties, gained any recognition at all. In the new regime technological progress was paramount and his pioneering work was embraced and celebrated. Though it would be some time before the rest of the world would learn of his insights, the government raised him to the status of a patron saint: He was awarded a lifelong pension, his complete writings were published, he was accepted in the highest scientific circles, and he passed the fifteen remaining years of his life honored as the "Father of Cosmonautics," ruminating quietly in his little home in Kaluga.

In the United States political events also affected Goddard's work. By 1917 the antiwar sentiment that had placed Woodrow Wilson in the White House had evaporated, and in June General "Black Jack" Pershing led the first U.S. troops overseas. This twist of fate ultimately inspired the Army Signal Corps to make Robert Goddard a proposal: If the Army financed his work, would he be willing to investigate the use of rockets as weapons? Goddard agreed and soon found himself up on the wind-wracked slopes of the Mount Wilson Solar Observatory overlooking the flat expanse of Pasadena, California, with a $20,000 budget at his disposal.

He immediately applied the money toward the work he had already done back East, and produced his first weapon the following year in the form of a recoilless rocket launcher. With this gun, he informed the army during a field test, a soldier could fire a blistering, eight-pound missile that could travel three-quarters of a mile from right off his shoulder with no more difficulty than squeezing off a round from his rifle.

The signal corps was so impressed by this invention that they immediately asked Goddard to develop a six-inch, high-velocity rocket that could be fired from an airplane. But four days later, on November 11, 1918, Germany surrendered and not long afterwards, Goddard found himself on his way back to Worcester on an eastbound train. His little recoilless launcher notwithstanding, he could see that progress with his rocketry work would come neither easily nor swiftly. Even with the support of the Smithsonian and the signal corps, he had made practically no headway toward anything that could be called a spaceship or even an atmospheric probe. Under the circumstances, he felt that it would be best to downplay his rocketry work and stay clear of any publicity, at least until he had a better handle on the engineering problems he needed to resolve.

Dr. Arthur G. Webster, Goddard's mentor and the professor who had overseen his doctoral work, didn't agree, however. Webster had inherited Clark University's physics department from the legendary A. A. Michelson—the first American scientist to win a Nobel Prize—and he believed that it was high time Goddard go public with his research. Goddard was appalled but Webster insisted: If Goddard didn't print the paper he had written for the Smithsonian, Webster would himself see that it was published.

Reluctantly Goddard agreed to reconsider and looked the paper over again. It wasn't as bad as he had thought. He had been circumspect about his wilder notions, though at the end of the article he *had* mentioned the possibility of landing a rocket on the moon, simply to demonstrate the possibilities of spaceflight. It was a casual comment, he felt, and nothing anybody would notice.

The Smithsonian agreed to publish the paper, believing a little publicity might draw funding for future research, and Goddard received his copy on January 3, 1920—a slim, simple volume bound in brown paper. The remaining 1,750 other copies were laid in the Institution's library, which, Goddard dearly hoped, would ensure their obscurity. For eight days it did. Then on January 12 the Yankee professor suddenly found himself, and his ideas, plastered across every front page in America:

MODERN JULES VERNE INVENTS ROCKET TO REACH MOON

That was the banner headline in the *Boston American.* Other headlines ran from the relatively accurate AIM TO REACH MOON WITH NEW ROCKET *(The New York Times)* to the utterly ludicrous SAVANT INVENTS ROCKET WHICH WILL HIT THE MOON *(San Francisco Examiner).* Goddard was appalled. Overnight he had been transformed from an obscure scientist into a wild-eyed maniac whom the media now called "the moon man."

Americans were equally astonished, but for different reasons. A rocket that could go to the moon in 1920 would be like announcing today that scientists had found an inexhaustible source of clean energy or cloned a human being. Cars and airplanes were exotic enough, but the idea of a machine that could roar up into the sky and then land on the moon was absolute madness.

Goddard read these headlines and wondered how this could have happened to him. He had, after all, only written a harmless scientific paper, one so laden with formulae and jargon that he couldn't imagine how newspapers from one end of the country to the other could possibly even know about it. As it turned out the Smithsonian itself had done him in by issuing a press release on January 11 that highlighted the very scenario Goddard thought he had so thoroughly buried. "An interesting speculation...," read the press statement, "is on the possibility of sending to the surface of the dark part of the new moon a sufficient amount of the most brilliant flash powder which, being ignited on impact, would be plainly visible in a powerful telescope."

Immediately after the articles appeared, Goddard was flooded with offers from all sorts of adventurers eager to leave the planet. More than one hundred intrepid souls volunteered for the lunar journey including a man by the name of Claude R. Collins, a captain in the Army Air Corps during World War I and the president of the Aviators' Club of Pennsylvania. He would not only fly to the moon, he wrote, he would travel on to Mars as well, so long as the good professor provided a $10,000 life insurance policy. A couple of days later a woman from Kansas City, who apparently had more faith in Goddard's engineering talents than Collins, announced that she'd go along, and didn't care if she was covered by a policy or not. A song was published with the enchanting title, "Oh, They're Going to Shoot a Rocket to the Moon, Love!" with the composer generously offering to turn over half his royalties to Goddard. No royalties, however, were forthcoming. And in Hollywood at the Mary Pickford Studios, an inspired press agent telegraphed the besieged Yankee physicist: WOULD BE GRATEFUL FOR OPPORTUNITY TO SEND MESSAGE TO MOON FROM MARY PICKFORD ON YOUR TORPEDO ROCKET WHEN IT STARTS.

Not everyone was so enthusiastic. *The New York Times* published a blistering editorial that criticized Goddard for his shoddy science. "That Professor Goddard with his 'chair' in Clark University and the countenancing of the Smithsonian Institution does not know the relation of action to reaction, and the need to have something better than a vacuum against which to react—to say that would be absurd. Of course he only seems to lack the knowledge ladled out daily in high schools . . ." But it was the editorial writer who had his physics wrong. Goddard had already proven that rockets would fly in a vacuum, but who would ever believe him now that he had been publicly humiliated in the pages of the august *New York Times?*

The irony, of course, was that Goddard had privately conceived far wilder notions than a flash-powder rocket to the moon. There were his fantasies about "operators" flying rockets controlled by gyroscopes and side jets, communication with extraterrestrial beings, manned journeys to Mars, ion propulsion (he even had a patent on this); but here he was being reduced to a buffoon by this one small afterthought that he had considered so harmless. He later wrote in his diary that he would have been better off concocting some wild scenario about a mission to Mars, a story that the press would undoubtedly have ignored as too outrageous.

Goddard made a few futile efforts to clarify his theories in scientific journals and various newspapers, but he soon gave up and retreated to his physics shop on the Clark campus. In time editors relegated "the moon man" to the back pages of their papers, then forgot him altogether, and finally Goddard's lunar antics disappeared into a strange cocktail of roaring-twenties barnstorming pilots and Ivy League boola boola. In the meantime, however, news of Goddard's moon rocket drifted overseas where it attracted the attention of an equally eccentric mathematics teacher from, of all places, Transylvania.

Suggestions of ways of
using the Jet — Solar energy Profits

1. Jet. (a) Have liquid fuel lead into an impossible combustion chamber, C,
(of course it would have to be formed in) using H₂O₂
or gasoline + N₂O₅.

(b) The same, except the pressure is developed
(allowed to) all through the chamber. (necessitates
stronger walls.

(c) An explosive, eg smokeless powder, taken
as the, or arranged so as to be, slow burning (solid)
combined. (if too rapid, use passages

or (Test this with   telescope
pointed towards the sun, in daytime — the
steels being painted black.

(Also try with   steel, on bed, with springs (called by
  — petals, or charges, so as in

(d) Remember, the next tier must be
set off when the pressure has fallen.
   "   have this grip the next smaller
     shell fitting into it.

or (2) have each crumble gunpowder ignite
upper tier, when all have been lowered
(by spring, & pressure has fallen)

(e) Of course means must be had to prevent
dangerous explosion if a molten penetrates.
Have v. H₂O₂, a wall (a), such that heat,
by ignite it till // swells up the spring.

(4) have such a mixture in c, that
such a thing will not explode, but
ignite, & when a charge will shut down

2. Application of solar energy.
1. Heat the jet
2. Electrify the jet
3. Heat other mass
4. Electrify " "
5. Perhaps have the extra mass as H + O; D.
introduce these into the steam boiler (turbine)
as previously described, thus permitting extra
(1, may be illustrated as —
(In the case of 2 & 4, the boiler (
turbine & generator could be used.

The reflectors had best be arranged as follows.
If each ready its sun up, by a separate, smaller
— mirror, This arrangement is
— economize, in the heavy supports
necessary if the mirror is all in one
piece.
Separate mirrors (In the case of one, each shell might
have its own mirror

354.
Have matter electrically sent off, after being
exploded, at the focus of a paraboloid; the surface
of which is charged

Opposite: Sketches from Goddard's notebook, a place where he recorded many of his more visionary ideas.

Left: Hermann Oberth tinkering in his workshop in 1929.

Hermann Oberth was an angry-looking man with murderous black eyes and a great eagle-beak nose, but despite his draconian looks, he didn't have a sinister bone in his body. Like Goddard he was shy and quiet and just as smitten with the idea of space travel. So in May 1922 when he read of the American's work in a Heidelberg newspaper, he immediately wrote him a letter in English so fractured and in a tenor so humble and polite, it was almost sweet.

> Dear Sir,
>
> Already many years I work at the problem to pass over the atmosphere of our earth by means of a rocket. When I was now publishing the results of my examination and calculations I learned by the newspaper, that I am not alone in my inquiries and that you, dear Sir, have already done much important works at this sphere. In spite of my efforts; I did not succeed in getting your books about this subject.
>
> Therefore I beg you, dear Sir, to let them have me. At once after coming out of my work I will be honored to send it to you, for I think that only by common work of the scholars of all nations can be solved this great problem.
>
> Yours very truly,
> Hermann Oberth
> Stud. Math Heidelberg, Germany.

For his part Goddard didn't care for the letter. He had felt all of these years that rocketry had been his own private quest, a universe where he alone traveled, and now here was this communication. Who was this Oberth anyhow, and what did he mean he was publishing his own "examination and calculations"? Nevertheless, he felt obliged to answer the letter, so he reluctantly sent a copy of his paper to Heidelberg and returned to working on development of a liquid-fuel pump.

The truth is that by the time he had written to Goddard, Oberth had already been thinking long and hard about spaceflight and rockets. During the war, in fact, he had proposed a design for an armed missile to the German high command. Not that he harbored any particular love for weaponry, but he felt that a rocket bomb striking Picadilly Circus would be so horrifying that it would quickly bring an armistice. The idea had only come to him because, like Tsiolkovsky and Goddard, he had been fascinated with spaceflight since childhood. As a boy he too had picked up a copy of Jules Verne's *From the Earth to the Moon,* and was so mesmerized by it that he didn't put the book down, except to sleep, for three days. He knew that Verne's idea of shooting a ship to the moon from an enormous cannon would never work—the concussion would flatten anyone in the capsule. But he was utterly taken by the idea, and he decided that even if Verne hadn't exactly come up with a feasible method for traveling to the moon, he would.

Oberth subsequently schooled himself in mathematics, discovered Newton's third law of motion ("For each action, there is an equal and opposite reaction"), and, just as Tsiolkovsky and Goddard had before him, concluded that a rocket was the only sensible way to leave Earth. By the time he had sent his letter off to Goddard, he had already solidified his ideas on rocketry in a long scientific paper he entitled *Die Rakete zu den Planetenräumen (By Rocket to Planetary Space).* He had hoped his professors at the University of Heidelberg would accept *Die Rakete* as a doctoral thesis, but he couldn't find an academic anywhere who would support it.

Discouraged, Oberth scrapped his scholastic hopes and published *Die Rakete* on his own, and to his complete surprise it sold out immediately. A second printing was issued and it sold out before the publisher could even get a third printing into the bookstores. The pamphlet was nothing fancy, but its opening caught the reader's attention with a short list of astounding statements:

> These I wish to prove in this book.
> 1. At the present level of science and technology it is possible to build machines which can climb higher than the earth's atmosphere.
> 2. With further improvements these machines can reach such speed that if left in space they will not fall back to the earth but will be able to resist the pull of gravity.
> 3. These machines can be built so that men can go up in them (probably without danger to their bodies).
> 4. It might be worthwhile to manufacture these machines when economic conditions improve, perhaps in a few decades.

Oberth divided *Die Rakete* into three distinct parts. The first section was filled with equations and dense discussions on

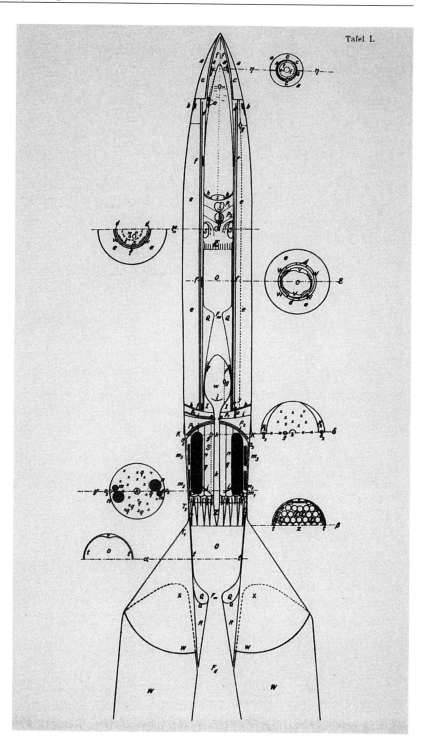

Oberth's plan for the Model B Rocket published in the 1923 *Die Rakete zu den Planetenräumen (By Rocket to Planetary Space)*. The Model B is a two-stage liquid fuel rocket. It was designed to perform geophysical research in the upper atmosphere, but was never built. A mixture of oxygen, alcohol, and water fuels the first stage. The second stage is fueled by liquid hydrogen and liquid oxygen.

regenerative cooling, trajectories, and escape velocities—the basics of rocketry. Section two outlined what Oberth called his Model B rocket, a two-stage contrivance powered by combining liquid hydrogen and oxygen. But part three was the section that captured the imaginations of his readers. Here he described an enormous rocket, the Model E, which he wrote would be capable of carrying humans to other worlds. This chapter also addressed the problems of spaceflight and furnished several creative solutions. The ship's pilot would use periscopes to view the Earth and stars; hooks and straps would keep the crew from floating weightless around the ship's interior, and space travelers would be outfitted in specially pressurized suits, with tanks of liquid oxygen and nitrogen supplying an artificial air supply.

Oberth quickly sent Goddard a copy of his book, and the Yankee physicist was stunned again. Even through the haze of the German text, Goddard could see that Oberth was wrestling with the very same theoretical issues that he was. But rather than keeping his futuristic thoughts to himself, the man was publishing them for everyone to read. Goddard was suspicious.

For a short time after he read *Die Rakete,* he engaged in a perfunctory correspondence with the person he would forever refer to as "that German Oberth," but eventually he cut off all contact. Years later he came to believe that Oberth had stolen his ideas and put them in his book, although there was no hard evidence that any such thing had occurred. Nor did Goddard ever address Oberth's work head-on either personally or in other scientific papers. Instead he merely stewed, scribbled more visionary notes in his diaries, and sent more papers laden with futuristic scenarios to Dr. Charles Abbot, his patron at the Smithsonian. His rocketry work would continue over the coming years, steadily gaining support, but aside from Abbot no one ever knew about his most revolutionary ideas, and no one ever would know . . . until after Goddard's death in 1945.

The fantastic nature of *Die Rakete* aside, another reason for its popularity was that it tapped a nerve in postwar Germany. The reparations demanded by the Allies following the war had decimated and humiliated the country, and a current of despair ran beneath the easygoing cabaret culture that had become so popular in the 1920s. Enormous posters around Berlin announced: "BERLIN, SEIN TÄNZER IS DER TOD." (Berlin, your dancing partner is Death.")

Germans longed for a new start, and rockets, it turned out, became a symbol of a hopeful future.

The man who eventually came to embody the possibilities of this new hope in the public mind was a huckster with a Homeric imagination named Max Valier. During the war Valier had been an aviator, and afterwards had made a name for himself as an expert and popular lecturer on everything from extravagant theories on cosmogony to fantastic scenarios about spaceflight. When he first read *Die Rakete* he was struck by its realism, and quickly saw that a less scientific version could reach an even larger audience, so he sought out Oberth and asked if the professor would be willing to collaborate on a revised edition. Oberth had already considered writing a more popular version himself because he knew that doing so would improve his chances of finding the funders he needed to design and build a real rocket. He gave Valier his blessing, and while the book was being rewritten, he went back to teaching high school in Transylvania.

Valier published the new book in 1924 under the title, *Der Vorstoss in den Weltenraum (The Advance into Space),* and it was an immediate hit. Not that the book was a sparkling piece of literature—it wasn't—but once again it tapped into postwar Germany's need for escape. The adventure of space exploration emerged as somehow imminent—as if the time when Germans could line up, board ships, and take off in droves for various parts of the solar system was just around the bend. Many who read the book, in fact, thought the sketches inside weren't visions of what *could* be, but renditions of actual rockets in the final stages of construction.

Imminent or not, the idea of space exploration now began to draw out a core of hardened enthusiasts. Several groups came in and out of existence in Germany, but the strongest of them emerged in July 1927 after Valier himself suggested that a society be formed exclusively to raise money to build and test real hardware. The group called their organization the Verein für Raumschiffahrt (Rocket Society)—the VfR. "The purpose of the union," read the original charter, "will be that out of small projects, large spacecraft can be developed which themselves can be ultimately developed by their pilots and sent to the stars."

At about the same time, space societies began to appear outside of Germany as well, cropping up in Hungary, France, Britain, and the United States, even as far away as South America. None, however, set down roots as deeply as those in the newly formed Union of Soviet Socialist Republics. There two very influential groups evolved—one in Leningrad and the other in Moscow, both cultivated by a firebrand engineer named Fridrikh Tsander, who had become fascinated with space exploration after reading Tsiolkovsky's first paper nearly twenty years earlier. Over the years Tsander had earned a reputation for being a first-rate thinker and engineer, but he had also worked feverishly to get the word out about the possibilities of space exploration, giving lectures anywhere he could find an open room and a few willing listeners. One evening in 1920 while speaking to a small group in Moscow he was startled to see the composed visage of Vladimir Ilyich Lenin in the audience, the very man who had directed the overthrow of an entire empire, and after the lecture Lenin summoned the young engineer to meet with him privately.

No one knows for certain why Lenin was interested in a subject as esoteric as interplanetary travel. Perhaps he simply liked the advanced technology that rocketry represented, or perhaps he had somehow sensed the political power space exploration would someday represent. Whatever the case, the meeting was prophetic because thirty-seven years later the launch of the first *Sputnik* satellite would stun the world, and, temporarily at least, symbolize the ascendancy of communism.

As rocketry eventually gained credence throughout the world, its unusual popularity in Germany continued unabated. Following the publication of *Der Vorstoss in den Weltenraum,* Max Valier, eager to encourage the public's interest and determined to raise funds for the VfR, decided to approach playboy and auto magnate Fritz von Opel. Cars were Opel's business—he had not the slightest interest in space exploration—but he could see that people were fascinated with rocketry, and that led him to wonder if he might not be able to capitalize on it somehow. So when Valier approached him, Opel asked one simple question: How long would it take to build a rocket car?

Valier suggested they talk immediately with a German businessman named Friedrich Sander (not to be confused with the Russian rocket scientist), who produced powerful signal and line-throwing rescue rockets for the Navy—simple, smokeless powder contraptions that Valier felt might work well enough on a car to draw a little more attention to rocketry. Sander's rockets weren't

as powerful or complicated as the liquid-fuel variety, but they were more reliable and less dangerous.

Shortly after this meeting, Valier had two Sander rockets attached to an Opel auto chassis and rolled the car out for a demonstration at the magnate's own headquarters at Rüsselsheim. Opel's own test driver Kurt Volckhart climbed into the vehicle and braced himself for the fury of the blast while Opel, Valier, and Sander looked on. The driver released the brake and hit the ignition button, the rockets kicked in and then—the rocket car of the future rolled down the test track at a miserable 9 miles per hour before gradually coasting to a standstill.

Opel was not impressed, and he said as much loudly. But Sander, who didn't care to have his rockets maligned, angrily grabbed a spare missile, strapped it to a long pole nearby, and lit the fuse. The pole immediately leapt into the sky, and Opel decided that further tests might be in order after all.

The next car was rigged with six rockets, and the one after that with eight. By the time the number reached twelve, the Opel rocket car was topping 70 miles an hour and attracting the attention of the press. Opel then built a long, sleek bullet of a machine he called the Opel Rak II, and mounted it with twenty-four Sander rockets. It was a true piece of high-tech automotive sculpture for its day, complete with an open cockpit, a little crescent windshield, and two stubby wings that sprouted from its sides to prevent the car from flipping over.

Next Opel arranged to run the car on May 23, 1928, at the Avus racetrack in Berlin, inviting two thousand guests from the cream of Weimar society, and arranging for his opening speech to be broadcast live over Berlin radio. When the rockets were ignited, the car performed flawlessly, hurtling across the racetrack at 125 miles per hour while the hushed and amazed crowd looked on. Instantly papers, magazines, and newsreels heralded the Opel Rak as irrefutable evidence that German science and technology were supreme. They speculated on rocket trains, planes, and ships, even rocket sleds and rocket bikes. There was talk about rocketing the mail to England and back, or possibly across the Atlantic to the United States. Products of all kinds suddenly assumed a propulsion motif: cigarette ads sported rockets headed for the moon, emblazoned with slogans like "Our age is marked by high-flying plans." Reporters came to christen all the hoopla *Raketenrummel*—rocket racket—and huge crowds gathered to

Max Valier, the king of *Raketenrummel*. Valier sensed the popular appeal of rocket power, and translated the works of rocket scientists like Oberth into writings that thrilled a broad audience. He also helped to found the Verien für Raumschiffart, or VfR, which came to include such distinguished members as Oberth, Wernher von Braun, Klaus Riedel, and Rudolf Nebel. Experiments like this liquid fuel rocket car were designed to draw attention to rockets and the VfR. On May 23, 1928, a car like this one reached speeds of 125 miles per hour on a Berlin racing track. Valier died in 1930 when a liquid fuel engine for a rocket car on which he was working exploded.

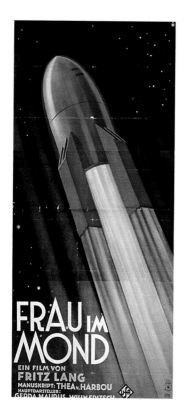

Poster from the 1929 Fritz Lang movie *Frau Im Mond*. Lang asked Oberth to serve as scientific advisor for the film and provided him with the chance to build a working rocket. The rocket was not completed for the movie, but the *Saturn V,* which launched astronauts to the moon forty years later, can trace its lineage to Oberth's efforts.

watch drivers of rocket vehicles of every kind skid, crash, and explode on fields, tracks, and pastures all over Germany.

Opel himself went on to build the Opel Rak III, and test it, unmanned, on a level stretch of railroad track where it topped 180 miles per hour. And Valier, who by this time had fallen out with Opel, raised the money to develop his own rocket car. But one day in May 1930, while he was working on a new liquid-fuel engine, the motor exploded, driving a piece of shattered steel through his aorta. He bled to death before anyone could do anything to save him.

All of the *Raketenrummel* failed to impress true space enthusiasts, particularly those at the VfR, who knew that powder rockets were fine for stunts but not for leaving the planet. That sort of undertaking would require liquid-fuel rockets and substantial funds to support their design and construction, and in an economy that had been decimated by the war and the reparations that followed, very little money was available. The future did not look promising. Then Fritz Lang became involved in the cause and events took another strange and unpredictable direction.

Fritz Lang was the high priest of German silent-film makers. He believed that films, properly made, tapped a primal wellspring in all people just as the old folktales and ancient myths once had. So when Oberth's *Die Rakete* came to his attention, he didn't see a technical book about rockets; he saw a mythic adventure about traveling to other worlds. The result was a screenplay which he brainstormed with his novelist/screenwriter wife Thea von Harbou: an epic space opera entitled *Die Frau Im Mond (The Woman in the Moon).*

Despite their often fantastic settings, Lang's films were legendary for their realism. The massive sets of his science fiction classic *Metropolis* had been unlike anything ever seen before, and he meant to bring this same realism to his production of *Frau.* To do so, however, to make every detail absolutely correct, he would need an expert, a man who appreciated complex technology and understood rocket flight. So, like Max Valier had before him, Lang called on Hermann Oberth.

Oberth was still teaching in Transylvania when Lang contacted him in the fall of 1928 and summoned him to Berlin. They met at Lang's house, and over coffeecake the genius director explained to the genius theoretician what a marvelous spectacle *Frau* would be: fabulous sets with lunar peaks rising up like cathedral spires, tons of bleached sand for the lunar surface, a fantasy world of unspeakable power and realism. What he needed most from Oberth was an absolutely authentic model of the ship that would carry the film's fated crew to its lunar destination. Lang also needed Oberth's scientific help on issues like propulsion and trajectories. What, for example, would the take-off *really* look like. Oberth, Lang promised, would be well paid for his advice, and would be given a workshop and whatever help he needed to build the model.

Oberth didn't know what to make of the offer, although he suspected it would draw attention to his work and help finance rocketry in general. A second expert that Lang had invited to the meeting, a young writer named Willy Ley, on the other hand, had no problem whatsoever in seeing the possibilities in Lang's offer. Ley, who had recently published a best-selling book entitled *Fahrt ins Weltall (Trip into Space),* was only twenty years old, but one of the most influential members of the VfR. He knew that when a Fritz Lang movie was produced, millions watched. To Ley *Die Frau Im Mond* was a rocket pioneer's dream come true.

The studio that would distribute *Frau* was UFA, a film conglomerate that controlled 75 percent of German movie production. This led Ley to wonder, privately, if the giant corporation might be willing to fund development of a rocket that could actually fly. UFA was approached but balked, so Lang, certain that the publicity of an actual rocket launch timed to coincide with the movie's premiere would generate great publicity, volunteered to put up some of his own money if the studio would match it from the movie's advertising budget. UFA finally agreed, and Hermann Oberth, the theoretician and advisor, suddenly found himself face to face with the same engineering nightmares with which Goddard had already been wrestling for ten years.

Robert Goddard would have been amused, or possibly horrified, by the situation Oberth now confronted. The New England professor knew firsthand how difficult building a working, liquid-fuel rocket was. Only a few years earlier, on March 16, 1926, he had finally managed to launch his own, but though it was the first of its kind, it had hardly taken the world by storm, traveling a mere 184 feet out over his Aunt Effie's Massachusetts farm. Success had come painfully and far too slowly.

Oberth, nevertheless, happily entered into his new enterprise unfettered by this knowledge. He had wanted to build a real

rocket, and now he had his chance, except that his abilities as an engineer fell far short of his theoretical talents. For months he struggled in his workshop on the UFA lot, trying to turn sketches into working metal parts, laboring to solve in weeks problems that really required years.

HOW TO BUILD A ROCKET IN YOUR SPARE TIME

The rockets built by Goddard, Oberth, and other pioneers around the world in the 1920s meant tampering with technology that was a good deal more complex and dangerous than the powder rockets that people were familiar with in those days. All that was needed to launch a fireworks rocket was a fuse and a match and an exhaust nozzle, but liquid-fuel rockets required motors and pumps, cooling systems to keep them from exploding from the extreme heat they generated, and solid construction to prevent them from flying apart under the extreme pressures that their fuels created. Attacking these problems completely from scratch with almost no money, the early rocketeers somehow accomplished the ground-breaking research that eventually made spaceships possible.

At Raketenflugplatz, for example, the MIRAK rockets that Nebel, von Braun, and Riedel were trying to launch had a habit of exploding because their combustion chambers grew too hot. One of their solutions was to line the chambers with aluminum, beryllium, or molybdenum, and add flanges that strengthened as well as drew heat away from them. Another major breakthrough made by both Goddard and the Germans was to continually cool the chamber by running the supercold liquid oxygen, which helped power the rockets, through a jacket that surrounded it. (But if the liquid oxygen grew too warm, it would vaporize and become useless.)

There were also the issues of aerodynamics. The rocket had to be slender and pointed at its nose in order to minimize wind resistance. Pumps, which were necessary to insure a steady stream of fuel, had to be small and light so that they didn't unbalance the rocket, and yet strong and fast enough to keep it fueled and moving. Weight, most of which consisted of the rocket's fuel, had to be properly distributed in separate chambers so that the rocket stayed on course both when it was

Robert H. Goddard poses with the world's first working liquid fuel rocket. The rocket was launched at the farm of his Aunt Effie Ward outside Worcester, Massachusetts, on March 16, 1926.

filled with fuel on take-off and drained of fuel during its flight. The ultimate solution to this problem became what Goddard called the "step rocket" or multiple-stage rocket, which used the largest chamber of fuel to get the missile off the ground and then successively smaller ones to keep the payload flying. As each chamber or stage ran out of fuel, it was dropped from the rocket and the next stage took over. American rocket expert David Lasser called this one of the most fundamental developments in the history of rocketry—without it space travel would never have been accomplished as early as it was.

The invention of the rocket really required building and designing many new machines that could work together in harmony. What is amazing, however, is that so many fundamental problems were solved in workshops that were little more than garages: Goddard's physics shop at Clark University and later his laboratory in Roswell, New Mexico; Oberth's back-lot lab at UFA; and Nebel's rundown barracks at the Rakentenflugplatz.

Oberth eventually realized that he needed help, placed an ad in the paper, and interviewed several applicants. One of them was a small man with ferretlike features who marched into Oberth's office meticulously dressed and announced, "Name is Rudolf Nebel. Engineer with diploma, member of oldest Bavarian student corps, World War combat pilot with rank of lieutenant and eleven enemy planes to my credit." Nebel, Willy Ley later recalled, probably never shot a single enemy airplane down in his life, and the closest he had ever come to being an engineer was when he had worked as a kitchen appliance salesman after the war. But showing the inconsiderable pragmatism of an eccentric, Oberth hired the man on the spot.

Later a second assistant, a Russian aviation student named Alexander Sherskevsky who had been supporting himself by writing articles—not all of them accurate—for German aviation magazines, also came on board. He had been sent to Germany to study gliding by the new Soviet government, but had stayed longer than he was supposed to and now was afraid to return.

Thus did destiny assemble the team to midwife the birth of the Space Age: a bewildered theorist, a Prussian militarist, and a misplaced Bolshevik. Together they faced an impossible deadline

The rocket constructed for the movie *Frau Im Mond* stands at its gantry poised for take-off. It looked surprisingly similar to the space shuttle.

with the movie's premiere on October 15, 1929, only a few short weeks away.

The first order of business was to choose a rocket fuel. Oberth decided that liquid oxygen and gasoline would be the most efficient, and tested the two ingredients' compatibility by injecting a stream of gasoline into a container of liquid air. It was a dangerous combination. No one was seriously injured in the first explosion, which merely rocked the shop and blew out one of its windowpanes, but the second one, not long afterward, nearly destroyed Oberth's vision in one eye, and shook him so thoroughly that it took a week before he could gather up the courage to return to his work. By now he had fired Shershevsky and in the face of the crushing deadline had drastically scaled back his original plans. He and Nebel would now cobble together a smaller, seven-foot rocket that would be loaded with sticks of carbon and liquid oxygen. When the fuse was lit, the plan went, the sticks would control the burn of the oxygen, which in turn would provide enough thrust to send the rocket on its way. Nebel managed to fashion an aluminum fuselage, but Oberth couldn't find any carbon sticks that would work, at which point he realized that the rocket couldn't be built. One night, with the premiere only a few days away, he left the shop and never returned.

UFA released a statement saying that the rocket launch would be canceled due to inclement weather, and the eccentricities of the rocket's complex machinery. The premiere, however, would go on as scheduled—which it did, without a hitch.

Even for a silent film, *Die Frau Im Mond* was overly melodramatic, but Oberth's model spaceship looked spectacular. In the take-off scene, Lang had it hauled from its hanger into a blaze of arc lights, a massive tool for taking humans off the planet and onto the moon. An enthralled crowd cheers as the ship rolls slowly toward its gantry, its design amazingly reminiscent of the space shuttles of today. Being the director that he was, Lang couldn't allow the rocket to simply hurtle off into the night without administering one last magnificent dose of high drama: the countdown, a piece of cinematic handiwork that survives today, and still works. It was so effective then that even without sound it's as though you can hear the engines roaring and the crowd gasping when the rocket bolts into the sky.

Oberth, Nebel, and Ley survived the experience of *Die Frau Im*

Actors in *Frau Im Mond* after their arrival on the "moon." No space suits were required.

Mond, but not without being seriously shaken and badly embarrassed. Ley, however, characteristically refused to accept defeat. The VfR still existed, he pointed out, and had even recently brought its membership close to one thousand, its highest level yet. UFA had magnanimously agreed to turn over the rocket hardware Oberth had built, free of charge, and from this the three pioneers hoped to salvage the rocket launch stand and various pieces of the more advanced motors and casings that had become casualties of the merciless deadline.

Together with the other founders of the VfR, they now decided they would redouble their efforts to build a working rocket. To do this, Ley knew they needed far more engineering expertise than they had, so he took it upon himself to draft fresh recruits. One of them was a full-fledged professional engineer who had recently joined the VfR named Klaus Riedel. Another was a strapping eighteen-year-old apprentice from the Borsig engineering works, an aristocrat whose father had been the minister of agriculture, something Ley thought might be helpful in the future. But more important than the student's connections was his long fascination with Hermann Oberth's writings and his genuine flair for engineering. Ley couldn't possibly have known then just how excellent his judgment would prove to be, because in this bright-eyed teenager he had chosen the man who would actually realize Hermann Oberth's wild dream of putting men on the moon.

He had recruited Wernher von Braun.

Top: Von Braun, still a teenager, with Rudolf Nebel carrying a model of the Repulsor rocket.

Bottom: The VfR August 5, 1930. From left to right: Rudolf Nebel, Franz Ritter, Hans Bermuller, Kurt Heinisch, unknown, Klaus Riedel, Oberth, and von Braun with the rocket that Oberth and Nebel had prepared for the film, *Frau Im Mond.*

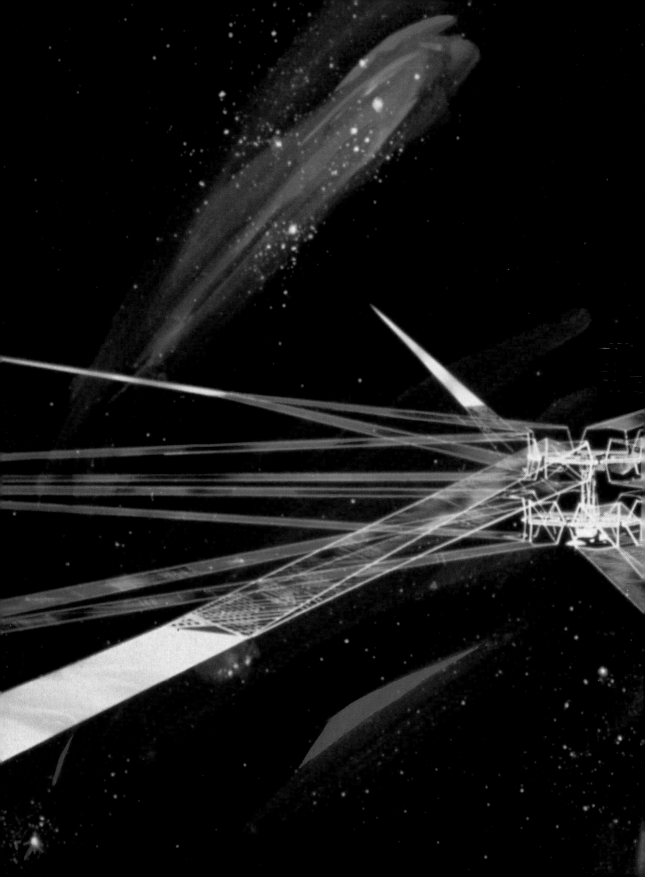

"Pragmatism always rests on the efforts of dreamers."
—Wernher von Braun

The Explorers

Above: Wernher von Braun and his brother Magnus with Private First Class Frederick P. Schneikert following their surrender on May 2, 1945. Von Braun had broken his arm in a car accident during the rocket team's exodus south from Peenemünde.

Left: Three V-2 rockets at a training area in Germany, March 1944.

I n the final days of World War II the Allies had finally broken through the last defenses of the Werhmacht, Hitler's Third Reich was crumbling, and German rocket scientist Wernher von Braun and his boss, General Walter Dornberger, were sitting by a radio in an old alpine resort listening to Bruckner's funereal Seventh Symphony, wondering whether they were about to be taken prisoner or simply murdered outright.

On the surrounding slopes German soldiers and a few engineers were practicing what they had been calling "defensive maneuvers," keeping their eyes out for any unwanted attackers who might be heading up the road or through the surrounding passes. But it was not American troops they were preparing to fight; they were on the lookout for their own countrymen—SS storm troopers under the command of Obergruppenführer Hans Kammler—an utterly ruthless man who would sooner gun them all down than allow the Allies to capture the scientific team that had designed and built wartime Germany's most important invention.

The ingenious machine behind all of this intrigue was the A-4 rocket, a lethal piece of engineering that Hitler had renamed the *Vergeltungswaffe* ("Vengeance Weapon"), or V-2,[2] and then, in a last desperate effort to reverse the course of the war, had rushed into production and dropped by the thousands upon the terrorized cities of London and Antwerp.

Their connection with the A-4 notwithstanding, von Braun

---

[2]   Vengeance Weapon-1 was the Fi-103, better known as the buzz bomb, a subsonic, pilotless mini-jet also loaded with a ton of the explosive Amatol that Germany started lobbing over the English Channel in the middle of June 1944.

and Dornberger had been doing everything in their power to be captured by the American army, but without success. Four months earlier, when the entire rocket team was still designing new missiles at Germany's secret Peenemünde firing range along the shores of the Baltic Sea, von Braun foresaw the fall of the Third Reich. When that fall came he knew that the members of his rocket team would be in grave danger, so he gathered the team together and asked them what they wanted to do. Every scientist and engineer voted to surrender to the Americans. As one of them explained later, they had no choice; the Russians terrified them, they abhorred the French, and they didn't think Britain could afford them. In America they might at least have the chance to carry on their work.

But arranging to surrender to the American army was easier said than done. Peenemünde was in northeastern Germany, a sector that would soon fall under the control of the Soviet army. American troops were hundreds of miles to the south moving in from Western Europe; the challenge would be to get there before the Soviets moved in or their own army tried to stop them somewhere in between.

A month after the meeting, von Braun commandeered a train, and with forged papers transported 525 of Peenemünde's scientists plus their families past the SS and into Bavaria where he hoped to find the Americans. At the same time he smuggled out thirteen years' worth of V-2 designs, which he later hid in a mine shaft outside of the small town of Bleicherode, 150 miles southwest of Berlin.

While von Braun and his team were laboring to be captured, intelligence experts in Washington, Paris, London, and Moscow were working just as hard to find them. They knew that a brilliant engineering team must have created the V-2, but had absolutely no idea who these wizards were or where they might be hiding. Then, one day, the team they had been searching out so assiduously suddenly turned up in the hands of an Army private from Sheboygan, Wisconsin.

Private First Class Frederick P. Schneikert had pulled guard duty with a couple of other GIs in April 1945 and was sitting in a damp culvert on the Austrian/German border when he noticed a man cycling down the side of the Alps, nicely dressed in a long leather coat, a clean shirt, and tie. Schneikert didn't like the looks of this. Rumors had been going around about German soldiers called Werewolves who were supposedly posing as GIs, infiltrating American units, and slitting the throats of unsuspecting troops. So he was especially cautious when he crawled up out of the culvert to take a closer look at the man on the bicycle. He flicked the safety off his M-1 and managed to recall some of the German he had learned from his grandmother back home and shouted, "Komm vorwärts mit die hände hoch!"

Suddenly the man jumped off his bike, walked forward with his hands up, just as he had been told, and began babbling in broken English that he was Magnus von Braun, brother of Wernher von Braun, the inventor of the V-2 rocket, and that he and the rest of the team who were hiding nearby wanted to surrender to the Americans immediately. In fact, it was absolutely imperative that he and his brother "see Ike as soon as possible." Schneikert was utterly baffled. He didn't have a clue as to who this madman was, and he certainly didn't know that he had just stumbled upon one of the greatest catches of the entire war.

The V-2 rocket was more a device of horrible psychological terror than a successful strategic weapon. Eighteen hundred had landed on Antwerp, and another nine hundred had struck London, killing a total of twenty-seven hundred civilians—an average of one death per bomb. Taken in the larger, strategic context of the war, they had failed; too few had been launched far too late. And yet they represented a remarkable piece of engineering. When American experts finally had the chance to look a V-2 over, they concluded that the Germans had been a good fifteen years ahead of them in the dark arts of ballistic rocketry.

Given the destructive power of the weapon, American intelligence officers might have assumed that a diabolical terrorist had conceived it. Instead they found themselves interrogating von Braun, a disarming, thirty-three-year-old with dark blond hair and a movie star's face. Like his father before him who had been the minister of agriculture in the old Weimar Republic, von Braun was a baron and a true Renaissance man. He had played the piano and cello as a child, and had even composed a few small pieces of music. He dabbled in several languages, and knew his way around ornithology and astronomy almost as well as he knew his way around a rocket engine. Von Braun told his interrogators that he had never intended to build weapons; he never believed it would come to that, but unpredictable events had led him into the missile-making business. His real interest was in building space-ships, and the truth was that he viewed the V-2 not so much as a supersonic bomb but as the incarnation of Hermann Oberth's "Rocket into Space," a machine that could take men to the moon.

There is no denying that Von Braun had long been enthralled with the idea of space travel. His mother was an accomplished amateur astronomer who had instilled in him an early fascination with the heavens. Later he was mesmerized by the *Raketenrummel* Max Valier had drummed up, and then, when he came across Oberth's book about space travel, he was entirely converted. But in a turn of events that largely set the course of the Space Age, his ascendancy as a rocket scientist had become strangely and irre-trievably linked to the megalomania of Adolf Hitler.

After the *Frau Im Mond* fiasco, Willy Ley had recruited von Braun as part of the VfR's renewed efforts to build working rock-ets, and by this time Oberth's ferretlike assistant, Rudolf Nebel, had moved the VfR's headquarters to a battered complex of cor-rugated buildings located on the grounds of an abandoned muni-tions dump outside of Berlin, a place Nebel grandly called Raketenflugplatz, or the Rocket Flying Place. Its major advantages were its open space and the run-down barracks where Nebel and his small group, mostly out-of-work engineers, could live rent free while they gathered scrap for rockets and dreamed of building spaceships.

Oberth, of course, had fled back to Transylvania where he could keep a better grip on his nerves, and Willy Ley would soon leave Germany as the Nazis rose to power. But Nebel stayed on, and together with his two new apprentices, Klaus Reidel and von Braun, had even managed to construct and launch a few working rockets.

Among them was a liquid-fuel contraption called MIRAK, short for minimum rocket, designed by Nebel, and the Repulsor, a small missile designed by Riedel. Neither was much larger than a good fireworks rocket, but the technology was considerably more complex. Still, no one was ever going to fly to the moon in a MIRAK. The VfR desperately needed money, but there were simply no investors willing to risk their capital in such an outland-ish venture. As Ley would put it later, the VfR had reached the point where it would be impossible for any club to build anything more than a few impressive toys, unless it was a millionaires' club.

The time had arrived for governments to intervene and, as it turned out, that is exactly what happened.

Following the *Raketenrummel* and the publicity that *Die Frau Im Mond* had generated, Colonel Karl Becker, a German Army Ordnance Corps officer and a renowned ballistics expert, had begun to wonder if rockets might not be useful as a new kind of artillery. He found them interesting because they skirted key provisions in the Versailles Treaty that had effectively eliminated weapons from the German arsenal as part of the agreement to end World War I. When the treaty was signed in 1919, rockets had been used mainly as flares and fireworks; no one perceived them as weapons. But now Becker suspected they might somehow be turned into long-range super bombs that could provide Germany with a powerful weapon without violating a letter of the treaty.

He assigned an artillery captain under his command, Walter Dornberger, to study the question. Dornberger was all business, a solid mechanical engineer who would prove to be a political in-fighter of Machiavellian stature, a talent that later greatly enhanced the success of the V-2. Dornberger's mandate was broad: Assemble a team to develop, quickly and secretly, some sort of liquid-fuel rocket weapon. But Dornberger found very little activity in this new field except among a small group of space enthusiasts living and working at a place they called Raketenflugplatz. In the spring of 1932 he paid a casual visit and quietly watched the rocketeers testing their small rockets. He was impressed enough to award them a contract for 1,000 reichsmarks (roughly $400) to build a Repulsor, and prepare it for a demonstration at the new military proving ground at Kummersdorf near Berlin. Von Braun, Nebel, and Riedel could hardly believe their good fortune. An opportunity like this could mean a government contract and enough money to build true rockets, not the little firecrackers they had been confined to playing with. They labored through the spring and by midsummer presented Dornberger with his Repulsor.

It took the better part of a day to set the rocket up for launch, but by afternoon they were ready. The fuse was lit and the Repulsor leapt two hundred feet into the sky, then suddenly shot off at a right angle toward the nearby woods and crashed into a stand of trees. Its parachute hadn't even had time to open.

Dornberger could see that he had his work cut out for him. No one, not even this group of diehard dreamers, was anywhere near building a missile that could deliver the punch he was after.

General Walter Dornberger. He hired von Braun and was in charge of the project that created the V-2.

The demonstration hadn't been a total loss, however, because, as Dornberger later recalled, he felt he had found something rare in "this tall, fair, young student with the broad massive chin." Although von Braun wasn't yet twenty, he obviously had a strong grip on the theoretical ins and outs of rocketry. During the demonstration, while Nebel struck Dornberger as more of a huckster than an engineer, von Braun seemed to be a problem-solver, willing to admit to flaws and prepared to attack them head-on. Dornberger never awarded a contract to the VfR itself, but three months after the Kummersdorf test, he hired von Braun to head up his rocket artillery unit. Nebel was not invited.

As talented as he believed von Braun to be, Dornberger also knew that a twenty-year-old student wouldn't be enough to carry out the mission that Colonel Becker had assigned him. He set out, therefore, with von Braun's assistance, to put together an all-star team of the few bright and creative engineers who had already done some serious rocketry work. Many of them, however, were leery of taking up with the military; designing weapons wasn't the same as designing spaceships or even rocket cars. Von Braun, on the other hand, argued that building rockets, even for the military, was simply a way of allowing the army to subsidize space exploration. No more scraping for funds or begging for parts; anyone who joined the team could settle into serious work, backed by serious money, and tackle solving the problems that serious space travel posed.

By 1934 Dornberger and von Braun had recruited the core of a rocket team, and the staff at Kummersdorf had ballooned from almost nothing to eighty. The money now rolled in, but the problem of building a rocket that could blaze into the sky and strike a target hundreds of miles away still remained. At this early stage even the most basic issues loomed enormous and forbidding. Designing the rocket's engine was only one of the problems. The engineers also had to figure out how to pump fuel into the rocket's chamber fast enough so that it wouldn't simply sputter out or blow up. Moreover, once the instrument was airborne, they needed to make certain that it headed in the direction they intended. After all, this wasn't a missile fired from a cannon and aimed like a bullet; it required stability and guidance—three-dimensional gyroscopic controls that could gauge where it was in space, and guide its thin metal exhaust vanes to prevent it from hurtling out of control and back to Earth like a piece of lead pipe thrown end over end. Radio links, an embryonic technology at the time, had to be designed to communicate with the missile, and all of this wiring and hardware had to be crammed somehow inside a very small space long before anyone had ever conceived of microchips or transistors. The only tools at hand in 1934 were (relatively speaking) big, clumsy servomotors, vacuum tubes, gears, and levers.

Robert Goddard could have told von Braun the frustrations they would soon encounter; he knew better than anyone that the basic, ground-breaking work was the hardest of all. By now the great aviator Charles Lindbergh had taken up Goddard's cause in the United States and had helped the professor land a grant from the Guggenheim family that enabled him to move his testing grounds to the desert outside of Roswell, New Mexico, where he could work in greater secrecy. There he had already managed to single-handedly invent gyroscopically controlled air vanes and multiple-combustion chambers, but even with the generous support of the Guggenheim family he had yet to launch so much as a ten-foot rocket up to the "extreme altitudes" he had written about nearly twenty years earlier. Nothing, in fact, had traveled more than a few thousand feet.

Unlike Goddard, however, von Braun was not a loner. He liked working with a team, and possessed an innate knack for management that allowed the Germans to make quick progress. He had an uncanny gift for slogging through quagmires of scientific data, literature, and eye-crossing technical drawings. He could talk to anyone on the team about any difficulty from thrust to aerodynamics, and could identify and translate the most important technical issues to the other members of the group without bogging them down in unnecessary detail.

He also developed an indispensable aptitude for handling the military bureaucracy. When the missile program was gearing up, the accountants had informed the team that they could order whatever equipment they needed to make rockets, but nothing unnecessary, like furniture or office supplies. Undeterred, von Braun and his team requisitioned "Appliance[s] for milling wooden dowels up to 10 millimeters in diameter," instead of pencil sharpeners, and rather than ordering typewriters they requested "Instrument[s] for recording test data with rotating roller as per sample." When the purchasing department rejected a requisition for a gold-plated mirror because of "insufficient justifica-

The Peenemünde assembly line.
When the war ended, the U.S. Army
moved in to Peenemünde and took
every scrap of worthwhile V-2
technology.

tion," von Braun told his assistant to file a new requisition with the explanation that they needed a gold-*plated* mirror because a *solid* gold one was too expensive. The requisition passed unquestioned.

Once the project was under way, progress came quickly, considering the hurdles that had to be overcome. The first rocket, dubbed the Aggregat-1, or A-1, was test-fired less than a year after von Braun began, and was immediately light-years ahead of anything the VfR had developed earlier. Powered by alcohol and liquid oxygen, it was four and a half feet long and generated a colossal 660 pounds of thrust. But it never flew because its nose was too heavy. By the following year, the A-2 was built with the rocket's gyroscopic stabilizer now repositioned closer to its center of gravity. Two models were ready for flight by December 1934— one called Max, the other Moritz—just in time to be hauled up to the North Sea island of Borkum as part of a demonstration for key members of the military high command.

By this time, Hitler had taken power in Germany and was flaunting the Versailles Treaty. Dornberger smelled an opportunity in this, and had begun to woo the nascent Luftwaffe, which was well funded under Hitler's friend Hermann Göring. If the A-2s performed well at Borkum, Dornberger hoped to attract Luftwaffe backing for his rocket team and move on to larger projects.

The A-2s didn't disappoint him: They roared up more than a mile high off their launchpads like great sleek firecrackers, and the Army and the Luftwaffe immediately began fighting one another for the privilege of pouring reichsmark after reichsmark into Dornberger's project. Given the added importance now attached to the program, a search began for a more secret site where truly large-scale and lethal ballistic missile tests could begin.

The place that was chosen lay on a gnarled stretch of land that separated the Bay of Stettin from the Baltic Sea near the village of Peenemünde, in the far northeastern corner of Germany. The rocket team moved in and its staff immediately swelled to three hundred people. As the personnel increased, a gathering of bizarre machines began to arise out of the sand, swamps and forests of Peenemünde: the world's largest wind tunnel, corrugated buildings crammed with people assembling giant rocket engines, and test stands with monstrous missiles strapped into them, bloated with deadly mixtures of alcohol and liquid oxygen, roaring away but going nowhere.

The first rocket built at Peenemünde was the A-3, and three of them were ready for launch by the fall of 1937. Twenty-one feet high and capable of thirty-six hundred pounds of thrust, the machines looked far more menacing than their smaller predecessors. But the inertial guidance system, which was supposed to ensure that the missile would fly straight and true, failed on all three, and the team found themselves face to face with a single problem so involved and delicate that it would take two hellish years for them to resolve.

As the work continued at Peenemünde, the Third Reich undertook the conquest of Central and Eastern Europe. In March 1938 Germany "absorbed" Austria, in September the Munich Pact awarded Hitler the Sudentenland, and in March 1939 the blitzkreig snatched up Czechoslovakia. Then Hitler's gaze settled on Poland, and war with Britain and France became imminent.

By now Dornberger had refined his vision of the new missile, and it was a tall order: He wanted a rocket that could deliver two thousand pounds of explosives to a target 160 miles away. It had to be slender enough to be transported by road or rail, weigh no more than twelve tons—eight tons of which would be the liquid oxygen and alcohol rumbling within it—and be able to develop twenty-five tons of thrust, nearly twenty times more than the A-3.

Six years of back-breaking research and experimentation were needed to resolve the monstrous technological problems involved in transforming the A-3 into the A-4. To generate the necessary thrust, for example, the new rocket would have to gulp fifty gallons of fuel per second, but no one had the slightest idea how to move that much fuel so rapidly. The problem was eventually solved when von Braun learned that local firefighters used pumps that shot water into hoses at high speed by centrifugal force, and an engineering genius named Walter Thiel, in charge of engine design, was able to build similar centrifugal pumps from scratch that could meet the new engine's demands. Special fins that could stabilize the rocket at supersonic speeds also had to be designed and built, an order that some engineers felt was impossible to fill; but with the help of Peenemünde's sophisticated wind tunnel, the impossible was accomplished.

The most frustrating problem was still the guidance system, the component that had failed on the A-3 and stubbornly resisted

Test launch of an A-4 (V-2) rocket at the Peenemünde firing range near the Baltic Sea. Wernher von Braun's mother had suggested Peenemünde as a potential test site.

resolution. After years of testing, the team finally solved the problem by placing a small preset clockwork inside the rocket's nose that turned the gyroscope to a predetermined position during flight, and shifted the rocket's vanes so that it would arc toward its intended target.

By the summer of 1942, the team was ready to test the first A-4s. The first two were successfully launched, but once they were airborne their engines mysteriously quit and they tumbled into the Baltic. By October 3 the team was ready to try again. Von Braun stood gazing at the rocket sitting beneath the noon sun—a great cigar-shaped metal cylinder anchored by four fins at its base. Not far off Hermann Oberth also stood waiting; at von Braun's request, he had recently been transferred from Dresden to a comfortable position at Peenemünde—a gesture of respect from the student to the master.

On ignition the A-4's engines roared, and the cigar slowly rose over the Baltic, so perfectly on course that Dornberger later recalled that it seemed to be running on rails. In less than half a minute it thundered through the sound barrier, the first missile ever to do so, and a minute after that it was gone. It covered 118 miles in five minutes and reached an altitude of sixty miles—the apron of space. That night Dornberger toasted the team. "Do you realize what we accomplished today?" he said. "Today the spaceship was born."

Despite the successful launch of the A-4, Hitler did not rush it into production. In fact, as far back as 1939, when he had toured a special demonstration at Kummersdorf, Hitler had been notably unimpressed with rockets in general despite the enthusiasm of many of his military commanders. During a static firing test, while the ground rumbled around him and everyone else looked on in amazement, Hitler hardly blinked. At lunch Dornberger sat down with the Führer to explain the potential of the rockets he had just seen, but Hitler, either incapable or unwilling to grasp what he had witnessed, simply went on eating his mixed vegetables and drinking his Fachingen mineral water.

Four years later, Hitler had become more indifferent than ever, partly because with U.S. and British bombers systematically destroying Germany's production capacity, and with supplies growing increasingly scarce, the A-4, and the team that made it, were looking more and more like expensive luxuries. Dornberger, however, argued that with the tide of the war turning against

Germany, the Third Reich needed the new weapon more than ever.

In May 1943, following an impressive demonstration of the A-4 for the high command, Dornberger pressed Albert Speer, the minister of armaments, to arrange another audience with the Führer. Speer apparently succeeded because on July 7 Dornberger and von Braun were summoned for a meeting at Wolf's Lair, Hitler's headquarters near Rastenburg in the mountains of East Prussia.

The meeting was ominous. Hitler, Dornberger later recalled, walked into the small screening room hunched over and looking pale as a corpse. The war was going badly for Germany; the Allies had mounted their D-Day invasion a month earlier and Allied troops were now sweeping east. The Führer threw off his enormous cape, sat down, and watched impassively as Dornberger rolled the presentation film and von Braun gave his lecture on the A-4. Afterwards Dornberger began to explain various aspects of the rocket and his plan for producing them, when something suddenly seemed to click in Hitler's head and he focused on what Dornberger was saying. How many of these weapons could be produced? he wanted to know. Nine hundred a month, Dornberger replied. Could that be increased to two thousand? No, there was not enough fuel on hand. Would it be possible to increase the explosive power of the missiles from one ton to ten tons? Again no, that would require an entirely new rocket. "But," said Hitler, "what I want is annihilation—annihilating effect!"

Going into the meeting both von Braun and Dornberger had been concerned that Hitler might expect too much. Now their worst fears had been realized: He expected the A-4 to instantly turn the tide of the war. Before the meeting ended Dornberger tried to explain to Hitler that the men who designed and built the rocket were sober engineers, and that no one had conceived of it as an annihilating weapon, but in mid-sentence Hitler turned on Dornberger, and screamed, "You! No, you didn't think of it, I know. But *I* did!"

Following that meeting, money, manpower, and materials flowed like water. Ministries tripped over themselves to make certain that the master builders at Peenemünde got whatever they desired. It didn't matter that von Braun and Dornberger downplayed their rocket, newly rechristened the V-2; Hitler had come to perceive it as the magic wand with which he would vanquish his enemies. Once it was loosed, he said, Europe—humanity itself —would no longer be able to endure any war.

The first V-2s thundered toward Western Europe on September 7, 1944, fourteen months after Hitler had ordered them into production. When the missiles began to explode on the streets and lanes of London, their arrival came almost as a relief to the Allies, because as it turned out, they were not the apocalyptic weapons that intelligence reports the previous year had described. Nor did they reverse the tide of the war or unleash the firestorm of might and vengeance that Hitler, in his desperation, had hoped they would.

What they did do, however, was dramatically demonstrate the powerful technology that von Braun's rocket team had wrought. The V-2s, spinning at supersonic speed into the hearts of ancient cities from launch points two hundred miles away, drove home the terrifying point that when you play with fire, the consequences can be unpredictable and horrible.

As Germany crumbled, the Allies commenced a mad dash to find von Braun's missile and the team that designed it. The Russians, it seemed, should have had first crack at the V-2 since the Peenemünde firing range fell within the designated Soviet zone in Germany. But when the Red Army arrived at Peenemünde in the summer of 1945, they found nothing more than a few scattered rocket parts, no sophisticated hardware, not a single knowledgeable engineer, no papers or blueprints—nothing at all that would help them build their own supersonic missiles. The U.S. Army, it turns out, had beat them on their own turf. Acting on information that von Braun had supplied, U.S. troops moved into Peenemünde and the V-2 production plant at Nordhausen, two hundred miles to the southwest, only a few weeks before the Red Army, and stole every scrap of worthwhile V-2 technology right out from under their noses.[3]

When the first American troops arrived at Nordhausen, they were dumbstruck. Among the great slabs of rock, stretching deep into a tunnel carved out of the side of a mountain, sat one rail car

Rubble following a direct hit by a V-2 in London near the end of the war.

---

[3]   After Normandy, the British carpet-bombed Peenemünde two times. Dornberger and others had suspected this might happen and had scattered production facilities throughout Germany to ensure V-2 production could not be wiped out in a single bombing raid. Thus, an assembly plant called Mittelwerk was created inside a mountain near the town of Nordhausen to protect it from bombing.

Top left: During the war, London's
underground subway stops became
ready-made bomb shelters.

Top right: London under aerial
attack, World War II. Though the V-2 s
didn't change the course of the war,
they raised the level of terror.

Right: British government official
checks the remains of a detonated
V-2.

after another loaded with great finned ships perched sleek and silent on their haunches. At the assembly plant itself they were greeted by the disturbing sight of hundreds of hollow-eyed slave laborers whom the Nazis had impressed to build the V-2s. Not far away, at the camp's barracks, thousands of bodies lay stacked and unburied, casualties of enforced labor. Toward the end of the war, an average of 150 workers had died per day, more, by some estimates, than all the people killed in London and Antwerp by the bombs themselves. Von Braun would never fully live down the controversy over just how much he knew about the slave labor at Nordhausen; throughout his career he was derided for making a Faustian bargain that countenanced the death of so many innocent people. But he denied any detailed knowledge of what was happening, and others have argued that most of the slaves had been transported from camps in Poland as the Allies closed in, and that those who died may have been killed by treatment they had received elsewhere, not from the work at Nordhausen. Years later von Braun told his friend, scientist and author Arthur C. Clarke, "I did not know what was going on, but I suspected. And in my position I could have found out, but I didn't and I despise myself for it."

In order to get its hands on as much V-2 technology as possible, the U.S. Army devised a plan to gather up the parts of one hundred rockets and ship them stateside where it could reassemble and test them. Troops rushed the bizarre parts and carcasses of the rockets onto trains, and the last missiles were shipped out only a few weeks before the Soviets moved in.

One of the first Russian experts brought in to survey what remained at Peenemünde was an engineer named Sergei Pavlovich Korolev, a stocky, tough-looking man with a broad forehead and intense eyes. Even with the scant evidence before him, he could see that amazing strides had been made at this remote firing range along the Baltic. At age thirty-eight Korolev was the Soviet's foremost expert in jet and rocket design and a man who had dreamed as long and passionately of space exploration as any member of von Braun's rocket team.[4] He had started out as a test pilot, some-

---

[4] Also on this expedition was Valentin Glushko, the man who would later design rocket engines that are still being used today. Glushko had worked with the great Fridrikh Tsander until Tsander's unexpected death in 1933, and he was thoroughly devoted to the idea of space travel. He was a big-chested scientist with a broad, serene face. It would be his huge rockets that launched *Sputnik* and Gagarin and Soyuz—strides that in the early days of the space race made the American rockets seem like so many misfired, small-fry roman candles.

Top: A young Sergei Pavlovich Korolev with his slide rule. Korolev was one of the Soviet experts sent to investigate what remained of V-2 technology at Peenemünde. He started out as a test pilot, but became one of the Soviet Union's top rocket scientists.

Bottom: Members of the team that designed and built the first Soviet liquid fuel rocket called the 09 or GIRD X. GIRD was an acronym for Group for the Study of Rocket Propulsion Systems, the organization in Moscow for which the team worked. Korolev (extreme left) directed the rocket's production.

thing you might sense if you saw him. He had a pilot's body— hard and compact, the sort you could cram into small places, and he had a test pilot's cool, unflappable demeanor.

Like Goddard, Oberth, and von Braun, Korolev had been inspired by the dream of space exploration early. After graduating from the elite Moscow Higher Technical School, he had worked at the Group for the Study of Rocket Propulsion Systems (GIRD), which had evolved out of Soviet versions of the VfR originated by Fridrikh Tsander. At age twenty-two, Korolev actually met Tsiol-kovsky, the "Father of Cosmonautics," and, the story goes, he was so impressed by the old visionary that he pledged to him that he would devote his entire life to making space travel a reality.

It was a pledge that carried weight because Korolev had an aptitude for succeeding at whatever he set out to accomplish. In 1932 he was placed in charge of GIRD's design and production department, and a year later launched Russia's first liquid-fuel rocket, the 09. Stalin, however, decreed that the main work at GIRD should be armaments, not space travel, and for four years Korolev went to work designing prototypes for jet planes, surface-to-air missiles, and big liquid-fuel rockets similar to the V-2.

However adept an engineer Korolev was, he made a poor Communist, and during Stalin's first purges in the 1930s he was arrested and sentenced to hard labor in the Siberian gold mines in Kolyma. Prisoners at Kolyma rarely survived their sentences, but a year later Korolev was saved when an old friend and col-league, Andrei Tupolev, who had also been arrested but subse-quently assigned to a top-secret engineering project in Moscow, requested that Korolev be allowed to join his program.

During the war Korolev's reputation grew, and though he was still officially a prisoner when he visited Peenemünde, he was placed in charge of reestablishing V-2 testing and production, a job that, a decade later, placed him in an ideal position to pull off one of the greatest feats of the Space Age.

THE DANGEROUS POLITICS OF ROCKETEERING: THE ARRESTS OF WERNHER VON BRAUN AND SERGEI KOROLEV

The careers of Sergei Korolev and Wernher von Braun

advanced rapidly because they were pioneers in an important and complex field. But because they also had the misfortune of working for two of the most dangerous political leaders in the twentieth century, they suffered catastrophic setbacks at the hands of the same governments that employed them.

Korolev barely escaped execution when his boss Marshall Tukhachevesky was arrested on suspicion of treason. Tukhachevesky was summarily executed on the night of his arrest, and because of his military work at GIRD, Korolev was found guilty by association and sent off to Siberia. At the time Korolev said that he expected to vanish "without a trace," but instead was saved by the intervention of Andrei Tupolev, one of the Soviet Union's great aircraft designers.

Korolev had always been open in his disdain of Stalin even as he worked for him, and waited until Stalin died in 1953 before he joined the Communist Party. He was never publicly recognized for his genius in the Soviet Union or elsewhere in the world until after his own death in 1966. After the launch of *Sputnik,* Khrushchev said that the identity of the man who had engineered the feat would not be revealed for fear that the West would attempt to kidnap or assassinate him. But a more plausible explanation is that Korolev's identity was kept secret so that it wouldn't dilute Khrushchev's own notoriety.

Von Braun was arrested on the evening of March 15, 1944, when the SS knocked on his door and took him, Klaus Riedel, and another engineer named Helmuth Grötrupp to a prison in nearby Stettin on charges of conspiring to sabotage the A-4 project. When Walter Dornberger learned of the arrest he was outraged and demanded an explanation. "Do you know," an officer told him, "that your 'closest colleagues' have stated in company . . . that it had never been their intention to make a weapon of war out of the rocket? That they had worked, under pressure from yourself, at the whole business of development only in order to obtain money for their experiments and the confirmation of their theories? That their objective all along has been space travel?" Dornberger didn't deny it, but said that such talk didn't constitute sabotage and that the charges were ridiculous. Furthermore, he said that losing von Braun and the two other engineers would slow the development of the A-4 and make its eventual deployment impossible. Several days later all three men were provisionally freed for three months, and eventually the case lapsed.

Did von Braun really intend to sabotage the A-4? Not a chance. Some historians believe he was arrested because he had gotten on the wrong side of SS Reichsführer Heinrich Himmler after Himmler had visited Peenemünde and, in a bid to take over the project, suggested to von Braun that he forsake Dornberger and work for him. Von Braun apparently made it very clear that he had no desire to have Himmler as a boss.

A year after the arrest all of these issues became moot because by then the entire rocket team had begun their search for the American army. Helmuth Grötropp, by the way, was the only key member of von Braun's team who elected to work for the Soviets. He remained in Germany after the war, and the Soviets offered him a nice house and a very high salary to resurrect V-2 production at Nordhausen. He later also worked in the Soviet Union, becoming the only rocket engineer to collaborate closely with both von Braun and Korolev.

"Either perish or overtake the advanced countries," Lenin had decreed in 1919. From the moment it came into existence the Soviet Union was a nation with a chip on its shoulder, determined to prove to the rest of the world that it could hold its own on whatever technological front it chose. In the years following World War II, Stalin's fear of the "advanced" countries, particularly the United States, reached a new and more ominous plateau.

On July 16, 1945, in a place called Jornada del Muerto (Journey of Death), a part of the Alamagordo bombing range in southern New Mexico, American scientists lit the fuse on a device they had nicknamed "Fat Man," the first atomic bomb. A searing flash bleached the desert, a ball of flame rose miles into the sky, and the future of the human race changed forever.

THE MANHATTAN PROJECT

Between 1942 and 1945 the United States brought together the brightest scientists in the country and spent $2 billion to produce the first atomic bomb. This undertaking, code-named the Manhattan Project, had its roots in a letter written by Albert Einstein to President Franklin Roosevelt in the summer of 1939 explaining the military potential of nuclear fission chain reactions and suggesting that it might be in the best interest of the United States to be the first nation to capitalize on this knowledge.

Roosevelt began the research in February 1940 with an initial funding of $6,000, but in 1942, when the United States entered the war, the project was placed in the hands of the War Department and eventually under the command of Brigadier General Leslie R. Groves. By this time there was a real concern that Germany would succeed in developing a bomb of its own. In the early 1940s in Berlin, Otto Hahn and Fritz Strassman, refining Enrico Fermi's earlier work in Italy, split a uranium atom into two fragments and liberated 200 million electron volts.

Scientists from Canada and Britain eventually joined the Manhattan Project, and the entire team went to work in a vacated boys' school on a huge, isolated mesa in the mountains of Los Alamos, New Mexico. When the first fissionable material was ready in the summer of 1945, General Groves had it delivered to New Mexico where the scientists had developed a way to make the plutonium into pure metal and fabricate it into an innovative imploding bomb. They called the first bomb "Fat Man" because of its squat, round shape. It was detonated at Trinity camp at Alamogordo on July 16, 1945, and unleashed the equivalent of between 15,000 and 20,000 tons of TNT, fusing the sand of the surrounding desert into glass and vaporizing the low metal tower in which it had been suspended. After the searing blast and the spectacle of the mushroom cloud, there was a moment of jubilation: The scientists had solved one of the most perplexing physics problems of their time. But resolving the theoretical complexities of nuclear fission created equally complex political and ethical problems.

Like Peenemünde, Los Alamos was the site of intense, expensive technological research funded by war. The Third Reich spent approximately the same amount of money to develop the V-2 as it had cost the United States to build the bomb. Ironically, it was the marriage of these two advanced and lethal technologies that essentially ignited the space race and made the first ICBMs and the first spaceships possible.

An intercontinental ballistic missile in its silo ready for launch.

> Luckily, no nation has yet fired an armed ICBM, but thousands of spaceships have been launched.

To Stalin Alamogordo represented a sudden and horrible inequity that he was determined to right as quickly as he could. The Soviet military had proven itself a powerful land army during the war, but it couldn't match the air power of the United States with its squadrons of long-range bombers and worldwide system of bases, and although it had three hundred divisions massed on the borders of Western Europe, they weren't worth a nickel against bombers armed with nuclear warheads.

Stalin conceived a simple counterplan: Develop a Soviet atomic bomb, and then find a long-range system that could deliver it. After considering various options, including a piloted orbital bomber, he decided to pour money into missile development and pushed Korolev to design a rocket with the power to cross continents. "Do you realize," he told his high command, "the tremendous strategic importance of machines of this sort? They could be an effective straitjacket for that very noisy shopkeeper Harry Truman. We must go ahead with it, comrades. The problem of the creation of the transatlantic rocket is of extreme importance to us."

The Soviets exploded an atomic bomb in August 1949, and solved the first half of the problem. Meanwhile Korolev continued improving and testing his rockets at a railhead at Tyuratum on the plains of Kazakhstan in an effort to resolve the second half. By 1952 he had concocted the T-2, a homegrown Intermediate Range Missile, and by 1954 he unveiled his R-7 booster, the Semyorka, the world's first true Intercontinental Ballistic Missile, an unholy union of the two most significant inventions of World War II: the atomic bomb and the rocket. This machine, he knew, would fulfill the wishes of the Soviet high command, but it could also become the spaceship that Tsiolkovsky had dreamed of fifty-six years earlier.

Korolev could afford to harbor hopes like this because Stalin had finally died in 1953, clearing the way for a nonmilitary, orbital satellite launch. He had already been using military rockets to perform scientific work for several years and, as early as the 1940s had launched Pobedas, his own variation of the V-2, to explore the upper atmosphere. By 1952 he was sending dogs to the very perimeter of the atmosphere to learn the effects of space travel on living things. Twenty-four canines had been outfitted with space suits, placed in the nose cones of various Pobedas rockets, and carried as high as their boosters would take them. Then a parachute would open and the dog would rock gently back to Earth while a little camera clicked away, recording the animal's reactions.

This nonmilitary work only served to reinforced Korolev's belief that Tsiolkovsky had been right—humans could safely explore beyond Earth. But to actually pursue a scientific launch he needed the backing of the new ruling politburo in Moscow, six men in the Kremlin who had taken over after Stalin's death, led by a clever political infighter named Nikita Khrushchev.

Shortly after taking power, the politburo had been stunned to learn that a missile development program even existed. The moment they found out, they summoned the engineer who directed it. Korolev marched in and proceeded to inform the new Soviet leaders that he had been working on machines that spit fire from their enormous engines, could fly over the North Pole, and could hurl down bolts of nuclear might on the West.

Of all the politburo members, Khrushchev alone saw that Korolev's rockets represented much more than fascinating technological novelties. He viewed them as a lusty political symbol—proof that the Soviet system worked, and that the U.S.S.R. was a power not only equal to the United States but superior to it. The trick would be to find a clever way to unveil for the rest of the world just what the Soviet Union had created without submitting it to nuclear war. As it turned out, Korolev's idea for launching a satellite was the perfect answer.

In the United States, the government had expressed no such abiding enthusiasm for satellite launches, but certain events had taken place that would lead to the same results. Immediately following the war von Braun and his vaunted rocket team were transported deep into the interior of the United States to the White Sands Missile Range in New Mexico, a godforsaken stretch of pan-flat desert less than one hundred miles north of El Paso, Texas, and not far from where Fat Man had been detonated. Members of the team later recalled they had never seen a place that looked less like Germany. During the day, the sun bounced off the sand and seared the eyes. Aeons of baking in the hot sun had made the Earth hard as a griddle, and no trees, not even scrub or cactus, clung to the barren landscape. It wasn't much to look at, but they had to admit it was an excellent site for testing missiles.

By 1946 the German scientists were successfully piecing the expropriated parts of the one hundred V-2 rockets that the Army had shipped from Nordhausen to White Sands, and they were turning them over at a steady rate for the military to test. Unaware of what was going on in Tyuratum, the Army brass would watch the V-2s arc into the cobalt blue sky over White Sands and measure distances and thrusts, and would scratch their chins, wondering how rockets might fit in with the military future of the United States. A few scientists from around the country were also invited to load the V-2s with cameras, Geiger counters, barometers, and whatever other instruments they felt might reveal the secrets of the upper atmosphere.

This idea of using rockets for scientific research was nothing new in the United States. Goddard had written about their scientific possibilities in "A Method of Reaching Extreme Altitudes" in 1916. But scientists had never worked with rockets that could send their devices eighty or one hundred miles into the sky, and they marveled at the new opportunities they represented.

Although von Braun wholeheartedly supported the scientific use of his rockets, he believed they had a much larger role to play. He intended for them to lead the *human* exploration of space just as Tsiolkovsky and Oberth and Goddard had dreamed. He still believed that more than anything it was the romantic desire to explore that drew humans to build these machines. But at White Sands he was making no tangible progress along these lines, and he was frustrated. When he came to the United States he had hoped to develop his A-9 and A-10 rockets, enormous spaceships that had been on the drawing board at Peenemünde, and which would be capable of ferrying material up above the planet to assemble space stations and spearhead travel to other planets. Instead he was relegated to managing the assembly of rockets he had invented years earlier, with no prospect for innovation.

However, the astronomers and physicists and upper-atmosphere experts—the purists at White Sands—looked down on von Braun's wild-eyed notions of interplanetary travel. Most liked him personally but mistrusted him because he was an engineer rather than a scientist, which to their way of thinking explained why he was interested in human exploration—it required big machines and big projects, exactly what engineers liked.

Cover of the March 22, 1952, issue of *Collier's* magazine showing an artist's conception of a spaceship rocketing away from Earth below.

Collier's

March 22, 1952 • Fifteen Cents

Man Will Conquer Space Soon

TOP SCIENTISTS TELL HOW IN 15 STARTLING PAGES

By the early fifties, the scientists at White Sands were also growing frustrated with their progress, but for different reasons than von Braun. Their tests had clearly demonstrated the scientific potential of rockets, but the effectiveness of their experiments was limited by the short time their instruments actually remained in space before they plummeted back to Earth. They longed for a missile powerful enough to launch their experiments into outer space and keep them there inside what Lloyd Berkner, a geophysicist with the Carnegie Institution in Washington, jokingly called an LPR, or a Long Playing Rocket. Unfortunately, no one had yet built such a machine.

Von Braun would have been happy to oblige, but the Army was not interested. He himself had only just embraced the idea of small satellites, accepting them as minilabs that might precede larger manned stations. By 1953 he had begun to push for the launch of a small satellite that could be carried aloft on an improved version of the V-2, a rocket known as the Redstone after the old Redstone arsenal in Huntsville, Alabama, where he and his team now worked.

In 1955 scientists throughout the world, particularly in the Soviet Union and the United States, began to focus on a project called the International Geophysical Year (IGY), a watershed event in the Space Age. The origin of the IGY dated back to 1950 when a young astrophysicist named James Van Allen, who had worked with the V-2s at White Sands, had invited Lloyd Berkner and a few other friends over to his house in Silver Spring, Maryland, for a little soiree. After talking about past international scientific efforts, all of them agreed that it might make sense to coordinate global high-altitude research with scientists from all over the world, and over the coming year their brainstorm eventually evolved into what became the International Geophysical Year. It was to be the first scientific effort aimed at developing a big picture of the way the entire planet worked, and as part of this effort participating nations were invited to launch an Earth-orbiting satellite, an invitation the United States accepted on July 29, 1955. A day later the U.S.S.R. also accepted, and from that moment on the two powers found themselves racing one another into space.

Korolev was overjoyed with the approval of his satellite plan, and quickly settled on a design that strapped sixteen additional engines to the base of his R-7 rocket's fuselage so that it would have enough power to haul what was now being called *sputnik zemli* (travel companion of Earth) into orbit. Korolev didn't delude himself that the satellite was anything very advanced. In fact, it was called PS for Preliminary Satellite and had the look of a strange toy, a silver sphere with four whiplash antennas that swept away from it. Inside was a simple radio transmitter that would begin beeping soon after it was launched. The beep would be flat and standard, like the beep a beacon buoy makes in the water so that you know where it is, which was precisely the idea: The beep would inform Soviet scientists that the little ball had in fact succeeded in circling the planet. Following a few Earth-bound preliminary tests, Korolev successfully test-fired two of his newly designed ICBMs out over the Kamchatka peninsula near Japan without incident. By August 1957 his spaceship was ready.

In the United States, however, success was more elusive. The U.S. satellite project seemed to be sinking beneath the weight of its own intricate design. Even though von Braun's Redstone rockets already provided a powerful first stage for a satellite launch, the Eisenhower administration had passed them over in favor of the Naval Research Laboratory's more complex Vanguard model rocket, which had been designed purely for civilian and scientific experiments. Eisenhower felt that it was imperative that the IGY satellite not be tarnished by the military. However, by the summer of 1957 all four of *Vanguard*'s static firing tests had failed, while the Soviets were continually hinting that the launch of *their* satellite was imminent.

By the end of September 1957 Walter Sullivan, science correspondent for *The New York Times,* was pretty certain that what the Soviets were saying was true. Sullivan was covering all aspects of the IGY and had come to Washington to attend a meeting on IGY rocket and satellite plans. The conference was held in the great domed interior of the National Academy of Sciences building beneath grand renaissance-style paintings illustrating man's efforts to comprehend the cosmos. In the spirit of the IGY the Academy had also set up a small demonstration of the Vanguard satellite in the library adjacent to the dome—a full-scale model of the Vanguard rocket and the satellite it would launch. At another display nearby, a tiny animated Vanguard capsule buzzed happily around the planet. This, the little display seemed to say, was the way it would look, after the launch.

Following days of endless descriptions and explanations

about Vanguard from American scientists at the conference, everyone was now eagerly awaiting any news the Soviets would have about their launch, but no details were forthcoming except the vague and ubiquitous message that it was . . . imminent. Still hoping to hear some news, however, Sullivan attended a small affair hosted by Russian scientists at the Soviet embassy on the eve of the last day of the conference. By this time Sullivan was so certain a Russian launch would happen soon that he had filed a front-page story saying as much for the *Times* that would be published on the next day, October 5, 1957.

All of the top American scientists came to the Russian party: Van Allen; William Pickering, head of the Jet Propulsion Laboratory in Pasadena; Herb Friedman, the naval research scientist whose experiments with V-2s had revealed the X-ray universe; Lloyd Berkner; and many others. Each hoped that even here at this social event Anatoli Blagonravov, the head of the Soviet delegation, would come forward with some morsel of information. But Blagonravov remained a block of granite, just as he had all week.

Sullivan had just arrived when a Soviet attaché told him he had a phone call. He excused himself, took the call in the lobby, and was informed by the news desk at the *Times* that they had just received a Reuters wire report saying the Soviets had launched a satellite called *Sputnik I* that was orbiting the planet every ninety-three minutes, beeping away at twenty and forty megacycles.

Sullivan hung up the phone, walked back upstairs, and surveyed the room. As he looked at the faces before him he realized that he was the only person who knew. No one else had a clue, including the Soviets themselves! They had been silent all week long because they didn't have any more information about the launch than anyone else. Either that or they knew when the attempt would be made, but had kept their mouths shut in case the satellite blew up on the launchpad.

After a moment's consideration, Sullivan walked up to a small cluster of Americans that consisted of Richard Porter of General Electric, Pickering, and Berkner, and dropped the news like a hand grenade. Berkner was the first to get his bearings. As the senior American scientist, he clapped his hands, asked for everyone's attention, and raised his glass to toast the Soviets on their accomplishment, the launch of the first man-made satellite in human history—the true dawn of the Space Age. The Soviets quickly shook off their surprise, and then, as Walter Sullivan later

The launch of *Sputnik,* October 4, 1957, stunned the American public and initiated an all-out space race. It was outfitted with two radio transmitters that emitted a continuous beeping, allowing the satellite to be tracked. It orbited the Earth every ninety-six minutes seventeen seconds for twenty-one days before burning up in the atmosphere.

recalled, they all smiled "like they had swallowed a thousand canaries." Meanwhile, just down the street at the National Academy, the little animated Vanguard—the tiny dream ship—whizzed around a fake Earth.

Sergei Korolev waited ninety-three minutes before *Sputnik I* had completed its first orbit and ground control could confirm that it was indeed overhead, beeping. Then he phoned Khrushchev at the Kremlin. It was around midnight in Moscow, and Khrushchev had stayed up awaiting the call. He congratulated Korolev and his team, and then went to bed.

The news, however, was not greeted with the same calm equanimity on the other side of the planet.

No one could have anticipated the media riot that followed *Sputnik* in the West—not Eisenhower nor Khrushchev, not the scientists, not anyone. The average American was truly stunned. How could the number-one nation on Earth be caught with its pants wrapped so tightly around its ankles? Yet there *Sputnik* was in orbit, the hard, silvery proof that the Soviet Union actually had bested the United States. Everything from American education to the free-enterprise system was called into question, and newspapers all across America compared the surprise launch to the appalling debacle at Pearl Harbor.

Senate Majority Leader Lyndon Johnson said that the Eisenhower administration had made one of the most monumental political and foreign policy blunders in the history of the nation. *Sputnik,* Johnson reminded America, represented the high ground, mastery of the heavens. Maybe it was all right with others in the government, he told reporters, but he for one didn't care to go to bed by the light of a Communist moon.

The panic was further exacerbated when details about the man who had orchestrated this magnificent celestial maneuver remained cloaked in secrecy. The West was only informed that the man behind *Sputnik* was known as the Chief Designer, and he came to be seen as a Soviet version of the Wizard of Oz, a dark and faceless figure behind the Iron Curtain who rotated the wheels and pulled the levers of the universe at his whim.

After seeing the world's reaction to *Sputnik,* Nikita Khrushchev immediately asked his wizard for another demonstration of Communist superiority. Four weeks later, on November 3, 1957, while Washington was still reeling from the first launch, Korolev

Vanguard or Rearguard? Two months after the launch of *Sputnik* the United States attempted to launch their own satellite. The rocket rose four feet and then exploded on the launchpad.

pulled another lever and launched not only a satellite, but a satellite carrying a dog named Laika inside, and the world was again astounded. The Soviets had launched a living animal into space and were monitoring the effects of the launch as well as radiation and weightlessness on the creature. The mission's success indicated that humans could probably survive orbital flight, which in turn implied that manned spaceflight was precisely the direction in which the Chief Designer was headed.

Through all of this, *Vanguard* sat at Cape Canaveral, landlocked. Officially it was still in its test stage, but after *Sputnik* the White House announced that the *Vanguard* launch on December 6 would *not* be a test, it would be America's official IGY launch. When the morning of the sixth arrived *Vanguard* sat shimmering in the winter sun by its gantry at Cape Canaveral. It was a beautiful piece of kinetic sculpture, four sleek stages stretching toward the sky. But the truth was that only the first stage had been successfully tested—the rest of it was a crap shoot.

As *Vanguard*'s engines ignited, a mighty flame roared against the launchpad and the rocket rose up four feet, then wavered and paused in time and space under an apocalyptic splash of fire. That was when the full horror of it sunk in: It was not going to fly. Then, very gently, bowing to Isaac Newton's law of gravity, *Vanguard* dropped back onto the launchpad and exploded in a thunderous ball of burning, errant fuel.

When the debacle was over, the launch crew went out to the pad to assess the damage. Blasted, charred metal was scattered everywhere, but to their amazement, among the wreckage they found the third stage laying intact, like a comatose animal. And from inside, the little satellite was innocently whistling away just as though it were in orbit.

More panic ensued. The press called *Vanguard* "Rearguard," "Kaputnik," and "Stayputnik." John McCormack, a member of the House Committee on Science and Astronautics, said that if the United States did not immediately overcome the Soviet Union in the space race, it faced nothing less than national extinction.

However, the United States did have a plan B, and it was already in the works on the day *Vanguard* exploded. At a dinner party on the evening of *Sputnik*'s launch, von Braun had cornered Neil McElroy, the man who was about to become the new Secretary of Defense, and told him that everyone already knew the Russians would succeed first, and that *Vanguard* was never going to make it. Why not give his rocket team a chance? McElroy agreed and granted von Braun permission to prepare a satellite launch as a backup to *Vanguard*. The rocket team hauled out the Redstone, now renamed the Jupiter-C rocket, and prepared a special fourth upper stage to hold the satellite itself. In truth, von Braun had already been working unofficially with Van Allen and Pickering to prepare instruments to fly on his Jupiter-C just in case *Vanguard* failed, and was already well on his way to having a rocket and payload standing by. The project was quickly sanctioned and the satellite was christened *Explorer I*. On January 31, 1958, it was launched into a flawless orbit, and von Braun became a national hero, the man who rose to meet the challenge of the Communist menace in space.

Even though *Explorer I* was not the first or even the second satellite in space, its instruments were far more sophisticated than those on the *Sputniks* and this, at least, led to one first for the United States. The payload had been equipped with an instrument designed by James Van Allen that detected radiation belts encircling the planet which are created when the Earth's magnetic field captures high-energy protons and electrons from the sun, a phenomenon that produces the Auroras at both poles. This was a major scientific discovery, the first, in fact, by a space probe. Most Americans didn't understand the scientific significance of the Van Allen Radiation Belts, but they did perceive in a vague way that the little grapefruits, as Khrushchev liked to call the United States' satellites, could out-perform their Russian counterparts if only U.S. rockets could manage to get them into space.

THE JET PROPULSION LABORATORY

Before the arrival of von Braun's V-2 rockets at New Mexico's White Sands Missile Range, the most advanced rocketry work in the United States was being done at the Jet Propulsion Laboratory in Pasadena, California, under the management of aeronautics genius Theodore von Karman and his protégé Frank J. Malina. JPL had evolved out of von Karman's aeronautical investigations at the California Institute of Technology in the late 1930s, and subsequent efforts to use jets to assist bombers into the air more rapidly, and to increase

William Pickering, James Van Allen, and Wernher von Braun hoist a model of *Explorer I*, the first successful U.S. satellite, in celebration on the night of its launch.

the speed of fighter aircraft. The Jet Assisted Take-Off, or JATO, project, helped the laboratory perfect its development of powerful solid-fuel rockets that operated differently from von Braun's (and Goddard's) liquid-fuel rockets. They later became the fuel of choice for many military missiles because unlike the more volatile mix of liquid oxygen and hydrogen, they could be placed in storage or silos without losing their effectiveness.

In the 1930s, when von Karman and Malina put their funding proposal together to form the JPL, von Karman decided to remove all references to rockets and rocketry because in the United States those terms had become so thoroughly linked with pulp science fiction. Instead he changed them all to the more euphemistic phrase "jet propulsion," which is how the institution got its name.

JPL became a large research center during World War II, and many of its solid-fuel rockets were later used to power military missiles deployed in Europe. *Explorer I,* the first American satellite, would never have successfully carried its payload into space if not for the work done at JPL. Although von Braun's Redstone was the first stage of the Jupiter-C spacecraft that launched it, the second, third, and fourth stages consisted of a battery of JPL's solid-propellant Sergeant missiles, and the preparation of the satellite's scientific payload was supervised by JPL's William Pickering.

Following *Explorer*'s launch, Pickering attempted to position JPL as the United States central agency for space exploration because of its strong scientific legacy—its expertise in rocketry and the emerging science of robotics, guidance, and communication, but the Eisenhower administration stopped that effort when it called for the formation of NASA in 1958.

Nevertheless, JPL, with 1,700 scientists and an additional staff of 2,500, remains one of the most powerful institutes under NASA's umbrella, responsible for all unmanned interplanetary exploration.

Still, despite a series of successful U.S. launches of increasingly sophisticated payloads, the sorcery of the Soviets' Chief Designer seemed invincible. When the United States announced plans to launch a satellite around the sun in March 1959, the Chief

Designer launched one in January. The Jet Propulsion Laboratory had plans on the drawing board to send a probe on a fly-by of the moon by July 1960; the Soviets crash-landed an unmanned *Luna 2* on the moon in September 1959, and then, just three weeks later, *Luna 3* successfully orbited the moon and returned the first photographs of its dark side. Military commanders pointed out that the unsettling truth about these missions was that they required extremely advanced guidance systems, the sort that could be used for the accurate delivery of nuclear warheads. The power of the Wizard was ominous indeed.

A maelstrom of scenarios suggesting the best ways to launch a cold war counterattack in the heavens emerged out of the rhetoric that followed the *Sputniks.* There were calls for a "missile czar" to handle the "missile gap"; and von Braun and the Army put together a plan that outlined a four-man experimental space station by 1962, a manned lunar expedition by 1966, a permanent moon base by 1973, and a manned expedition to a planet by 1977—all to be supported by a family of superrockets. A little later, the JPL, which was now beginning to specialize in unmanned interplanetary exploration, proposed a series of missions to the moon, Venus, and Mars with probes so sophisticated that they would take the rudimentary instrumentation of *Explorer I* right into the untapped and futuristic world of modern robotics.

    The Eisenhower administration, however, called a halt to these scenarios in 1958 when it created NASA, the National Aeronautics and Space Administration, a new branch of the government whose job would be to concoct a space policy out of the witches' brew of these plans, and spearhead the investigation of the final frontier. For the first time in human history, a single agency now existed exclusively for exploration beyond Earth, a situation that suddenly funneled billions of dollars directly toward the very wild-eyed dreams for which the old pioneers had been scoffed at. The irony was that NASA did not spring from the dreams themselves, although dreamers were involved; it rose up out of the cold war and the persistent belief that the United States must catch the Soviet Union or be defeated.

    Over the next few years NASA allayed some of these fears by designing and launching a series of impressive spacecraft like the Pioneer probes, *TIROS I,* the first weather satellite, and *Echo,* the first communications satellite. These successes didn't attract as much attention as NASA might have hoped, but they did show a

The seven Mercury astronauts. Mercury was America's first manned spaceflight program. Its astronauts were selected on April 9, 1959, only six months after NASA was formally established. Back row, left to right: Alan B. Shepard, Jr., Virgil I. Grissom, L. Gordon Cooper. Front row, left to right: Walter M. Schirra, Jr., Donald K. Slayton, John H. Glenn, Jr., and Scott Carpenter.

Top: Yuri Gagarin. On April 12, 1961, the Soviet astronaut became the first man in space. Gagarin orbited the Earth one time and became an international hero. Once in orbit, he radioed back, ". . . one can see everything."

Bottom: Khrushchev, Gagarin, and Gagarin's wife Valentina in a parade celebrating the astronaut's historic mission.

certain sane and workmanlike progress in new space applications. At the same time, Eisenhower had gone ahead and reluctantly okayed the Mercury Program, which was to place an American in space by 1961. But just when it looked as though the gap would soon be closed, the Chief Designer triumphed again on April 12, 1961, when Yuri Gagarin, a boyish-looking Russian test pilot, crawled into one of Korolev's ships, rode it around the Earth, and became the first human ever to leave the planet.

If anything, the newest round of panic was worse than the one following *Sputnik.* All of Congress seemed to feel that NASA had failed its mission while the Chief Designer had once again gleefully pulled the cosmic levers. In a hearing held the day after the launch by the House Committee on Space, Congressman King from Utah, asked NASA Associate Administrator Robert Seamans whether the Soviets, given their greatly advanced technology, were going to land men on the moon in 1967 to commemorate the fiftieth anniversary of the Red Revolution. Seamans allowed that the Soviet government hadn't let him in on all of their plans, but said that a NASA program called Apollo, which proposed to orbit or even land men on the moon at some point in the future, had been discussed in the Eisenhower White House, although it had never been taken seriously.

Well, King wanted to know, *could* we go by 1967? This question put Seamans in a difficult position because he did feel that it could be done, provided that Congress and the president and the American people were behind it one hundred percent, and were willing to spend a lot of money.

The problem with Seamans making this admission was that the Kennedy administration was in no way committed to going to the moon at the time. In fact, that same day Kennedy had held a press conference and all but abdicated the space race. "The news will be worse before it is better, and it will be some time before we catch up," he said. But the president quickly realized that surrendering the space race, especially in light of other recent debacles such as the Cuban Bay of Pigs invasion, carried a heavy political price. So within weeks Kennedy reversed his position and sent a one-page memo to Vice President Lyndon Johnson, which essentially said that he was tired of the Russians beating the United States in space and wanted to do something to prove to the world that the United States was not second best. Johnson in turn passed the request on to NASA and the Department of Defense,

THE WHITE HOUSE

WASHINGTON

April 20, 1961

MEMORANDUM FOR

VICE PRESIDENT

In accordance with our conversation I would like
for you as Chairman of the Space Council to be in charge of
making an overall survey of where we stand in space.

1. Do we have a chance of beating the Soviets by
   putting a laboratory in space, or by a trip
   around the moon, or by a rocket to land on the
   moon, or by a rocket to go to the moon and
   back with a man. Is there any other space
   program which promises dramatic results in
   which we could win?

2. How much additional would it cost?

3. Are we working 24 hours a day on existing
   programs. If not, why not? If not, will you
   make recommendations to me as to how
   work can be speeded up.

4. In building large boosters should we put out
   emphasis on nuclear, chemical or liquid fuel,
   or a combination of these three?

5. Are we making maximum effort? Are we
   achieving necessary results?

I have asked Jim Webb, Dr. Weisner, Secretary
McNamara and other responsible officials to cooperate with
you fully. I would appreciate a report on this at the
earliest possible moment.

Above: "...this nation should commit itself to achieving the goal before the decade is out of landing a man on the moon and returning him safely to Earth." Kennedy gave his speech before Congress calling for an all-out U.S. effort to send men to the moon on May 25, 1961. "No single space project of this period would be more impressive or more important," he said.

Left: The day after the Gagarin launch, President Kennedy held a press conference and all but abdicated the space race, but he quickly realized the political price he would pay, and sent this memo to Vice President Johnson, which resulted in a plan for landing astronauts on the moon.

and over the weekend of May 6 and 7, 1961, an old Department of Defense proposal for a manned moon landing was transformed into a decade-long master plan for space exploration.

On the previous Friday, May 5, before this transformation began, Alan B. Shepard was launched in a modified Redstone rocket from Cape Canaveral into the upper atmosphere from which he plummeted like a cannonball into the ocean 380 miles down range. It was not as spectacular as Gagarin's flight, and in fact Gagarin was later quoted as saying, "We've sent some dogs up and down like Shepard." Nonetheless, it raised national morale and brought American space exploration back within the realm of possibility.

On Monday morning the finished report was delivered to Lyndon Johnson just in time for the Rose Garden ceremony in which President Kennedy awarded Shepard the Distinguished Service Medal for being the first American in space. Like most reports in Washington, this particular one might easily have found its way into a file cabinet and quickly faded from memory. However, two weeks later, on May 25, Kennedy announced before a joint session of Congress, with all the fire and brimstone he could muster, precisely what the report had recommended, and thereby dramatically redirected the course of the Space Age.

"I believe," Kennedy announced, "that this nation should commit itself to achieving the goal before the decade is out of landing a man on the moon and returning him safely to Earth. No single space project of this period would be more impressive or more important to the long-range exploration of space; and none will be so difficult or expensive to accomplish."

Kennedy's speech amounted to a declaration of war, and it did for the U.S. space program what Hitler's ambitions had done for the development of the V-2, except more. The gauntlet was thrown down and the Speech of Abdication, made only a few weeks earlier, disappeared from memory.

NASA now became a money sink. The projected cost of Apollo was $20 billion, and NASA's budget ballooned from $1 billion to $6 billion per year. No price was too great to assert America's mastery of the heavens, and Kennedy's continued speeches on the subject had a way of couching the undertaking in language indicating that Apollo was more than science and more than politics—it was a noble human endeavor, a moral necessity—and that if anyone was going to undertake and accomplish such difficult and high-minded achievements it would be Americans.

In the new push for the moon, von Braun was assigned the job of building the rocket that would carry the first humans there. The Chief Designer, however, had far more momentum than von Braun, and ambitious plans of his own. The Russian manned space program, run under Korolev's steady hand, still remained well ahead of NASA's. On August 6, 1961, a second cosmonaut roared into orbit and hurtled around the Earth for an entire day, a full six and a half months before John Glenn made the first U.S. orbital flight. A year later another cosmonaut flew into orbit, followed by a second cosmonaut launched the very next day, a maneuver that placed two Russians into orbit simultaneously. NASA had successfully launched only one astronaut into orbit, now it seemed the Russians were launching whole squads of cosmonauts.

Every time Korolev struck he added diabolically creative touches like this double launch. Next it was a woman. Later came the first capsule with two cosmonauts, followed by a capsule with three. Then there was the first space walk and the docking of two ships in space. These were all known as the Voskhod missions and all of them were firsts. The Chief Designer's sorcery seemed endless. He had taken mystical command of the heavens, like the ancient Magi, and each new Soviet venture in space took on the aura of a modern astrology, with each new man-made object auguring a dire future for the United States.

THE SHIPS THAT NEVER FLEW

It isn't well known, but before the *Saturn V* rocket and the other components of the Apollo program were ever designed and built, plans for three very different spacecraft were also considered for the journey. One was a colossal ship called NOVA which would have towered 360 feet high—as large as a World War II light cruiser and twice as high as the Statue of Liberty—and would have made an express run directly from Cape Canaveral to the surface of the moon like the spacecraft seen by millions in the science fiction movies of the 1950s. Such a ship would have had to generate 12 million pounds of thrust, but it was questionable whether engines powerful enough to actually launch something so large could be built. In the end the answer turned out to be no.

The second plan called for the construction and launch of two advanced Saturn rockets, each still a monstrous 320 feet long, which would circle Earth, dock in orbit, and then depart for the moon. One would carry the fuel for the trip, the other the crew and supplies. This was von Braun's preferred plan and could have been mounted fourteen months earlier than the *Nova* program, something that was of paramount importance in the heat of the space race.

The third plan took an entirely different approach from either of the first two, eliminating the need to land a large rocket on the moon. It called instead for a ship to take three astronauts into orbit around the moon with a smaller four-legged lander attached that engineers called "the bug." Once the three astronauts reached lunar orbit, two would crawl into the bug and descend to the surface. A day of exploration, exaltation, and flag planting, and then the astronauts would blast off with an engine that required a paltry 30,000 pounds of thrust to escape lunar gravity and dock a hundred miles above the moon, where it would rejoin the mother ship and return to Earth. If this plan sounds familiar, it is because it was the one that was ultimately chosen, and succeeded in landing two astronauts on the moon on July 19, 1969.

If the truth were known, however, the Chief Designer's grip was beginning to slip. All of the recent firsts had actually been nothing more than a series of stunts that Khrushchev had forced on Korolev to impress the world, and although the United States didn't know it, they were draining Korolev's money and resources, and ultimately dealt a death blow to his lunar program. Two years before the Voskhod missions, in 1964, Korolev had already conceived Soyuz, a family of ships designed to link up in Earth orbit and then circle the moon with a crew of three. The rocketry that placed them in orbit would not represent any great technological leaps, but linking ships in space and then sending them out of Earth orbit would. Korolev believed he could have the system flying by 1965, and could launch a manned mission around the moon by 1966. If that succeeded, he would then be able to add a lunar descent module and orchestrate the first human landing on the moon. Korolev wanted to start developing the craft immediately, but Khrushchev couldn't abide spending time on research and development while the United States was busy mounting the second phase of its manned program, called Gemini. To Khrushchev, it didn't matter that one Soyuz mission three years later would place the Soviets far out in front again—the immediate political momentum would be lost. Thus Korolev was forced into the Voskhod missions, and his Soyuz designs were delayed four years while NASA began to pick up the pace.

Korolev's plans suffered in other ways as well. After the initial successes of the Luna moon probes, key weaknesses in the areas of instrumentation and guidance arose and the probes suddenly developed problems finding their destinations. The first three Soviet spacecraft to Mars never even made it to the Van Allen Belts, while the first Venus probe, *Venera I,* arced off into space and mysteriously died a little more than a million miles from Earth. Many rockets thundered off the launchpads at the Baikonur Cosmodrome in the plains of central Asia, but of all the probes dispatched to Mars, Venus, and the moon only one succeeded; the other seventeen became cosmic flotsam or exploded before they made it into orbit.

Even more trouble lay ahead for the Soviet program. In October 1964, Khrushchev was ousted, and then, a little more than a year later, Korolev died of a heart attack during an operation to treat colon cancer. It was a punishing blow to the Soviet space effort. No one man in the U.S.S.R. had combined so much passion and so many talents as brilliantly as Korolev had, and after his death the Soviet program seemed unable to regain its momentum.

Although Korolev's death finally brought him the recognition he deserved at home, his passing was unknown in the West. Even in 1966 the United States still seemed to be losing ground. Nevertheless, the money flowed, and technicians and scientists from Cape Canaveral to Caltech continued to work overtime hoping to win the race to the moon. Then finally the momentum seemed to shift. In one two-month period, four flawless Gemini missions were flown in orbit around Earth, while Soviet cosmonauts stood earthbound and flatfooted.

These, and the Apollo missions that followed, became sources of wonderment and comfort for a population faced with a decade that was growing increasingly bleak with race riots, as-

sassinations, and the Vietnam War. The nation seemed to be un-raveling, but the race for the moon, at least for some, was a torch in the darkness.

Support for the manned space program had also helped fi-nance a long and impressive march of robotic explorers. Break-throughs in miniaturization, microprocessing, robotics, and solar power enabled probes like the Rangers, Surveyors, Mariners, and Pioneers to scout the inner and outer solar system, and, despite many early failures, they hit their marks with increasing accuracy, unlike their Soviet counterparts. The cold war had also pushed the use of surveillance, weather, and communication satellites, which soon began to change the face of the Earth sciences, mete-orology, and international communications.

Despite Korolev's death, the Soviet space program pushed on. There were a couple of successes like the *Luna 9* and *Luna 10* probes, which successfully soft-landed on the moon and sent back the first pictures from the surface of a celestial body, but while these further rattled the West, the remainder of the Soviet program was crumbling. The N-1 rocket that was supposed to carry cosmonauts to the moon as part of Korolev's Soyuz program just wouldn't work. A monstrous machine 310 feet from tip to tail, it was powered by thirty engines theoretically capable of devel-oping 150 tons of thrust each, but it was far from a flawless piece of engineering.

The N-1, and all of its problems, had been born out of a split between Korolev and his longtime collaborator Valentin Glushko over the kind of fuel that should be used to power it. Glushko had finally resigned, and then following Korolev's death the rock-et's design was drawn and quartered among the various design bureaus in charge of its development. So while the American *Saturn V* rocket launched the first humans on a voyage around the moon in December 1968, the N-1 languished. In 1969, it was finally ignited for a test run, and blew up on the launchpad, as did the next three. Even if engineers in the United States didn't know it, Soviets engineers did: the U.S.S.R. had already lost the race to the moon.

The *Saturn V,* on the other hand, was a clear and unadulter-ated success. It was thirty-six stories tall, could generate 9 million pounds of thrust, and launch a 150-ton payload into orbit around the moon. Its bottom stage was so enormous it had to be shipped to Cape Canaveral by barge—no other earthly form of transporta-

The *Apollo 11* lander descends to the moon, carrying visitors from another world.

tion could handle it. To prevent rain clouds from forming above the roof of the building in which it was assembled, engineers had to design a special air-conditioning system for the workers inside.

With the Saturn rocket von Braun ascended to the pinnacle of his career; it was a magnificent spaceship. He had begun work on it in 1962 and husbanded it with his old rocket team through every stage of its design, construction, and testing. Despite being five times the size of anything that had ever flown before it, it performed eleven missions flawlessly. On July 16, 1969, with the moon beckoning overhead, the *Saturn V* rumbled on its launch-pad at Cape Kennedy and hurled *Apollo 11* at its target. Three days later a breathless world looked on while the lunar module circled the moon, its crew searching for a place to land. Not long afterwards the first human beings stepped out of their spaceship and walked on the surface of an alien world. Of all the firsts this was the one that mattered most.

Although few remember it, on the same day a Russian probe called *Luna 15* was also surveying the moon for a landing site. *Luna 15* was to touch down and return samples of the moon's soil—an ambitious mission, but under the circumstances little more than a last desperate effort to salvage some glory in the waning hours of the moon race. *Luna 15* attempted a landing in a place called the Sea of Crisis, and crashed.

As von Braun watched his *Saturn V* carry *Apollo 11* into the sky that night, he was already looking ahead. With the momentum of Apollo behind them, he felt it was time for NASA to push harder than ever for a human mission to Mars. Plans were soon submitted to Congress and the White House calling for the construction of a space station, the creation of a shuttle craft that would link the station with Earth, and a mission that would send the first humans to Mars. All of this, von Braun believed, could be accomplished by the early 1980s.

The nation, however, was in no mood to extend its horizons. A lot had changed since 1961 when Kennedy had implored Congress for funds and initiated the Apollo program. Even as the last five lunar missions took place, they seemed somehow anticlimac-tic—almost a chore, and after the calamities of the 1960s, spending billions of dollars on space stations and Mars missions seemed like a ludicrous waste of money.

NASA's budget plummeted from more than $20 billion at the height of the Apollo program in 1964 to a low of $6 billion by

1974. Even before the budget hit bottom, von Braun grew discouraged and left the agency to take a job with aircraft manufacturer Fairchild Industries in 1972; and the old rocket team, now getting on in years, began to dissolve. Then in 1977 von Braun died of colon cancer.

It was difficult to imagine a person more perfectly fashioned for the Space Age than Wernher von Braun. He was amazingly tough and single-minded, yet joyful in his work, and like Korolev he bridged the gap between the old visionaries and the modern Space Age practitioners. There aren't many people who dream up something as fantastic as a trip to the moon in their boyhood and then go on to become the central figure in making that dream come true. His ambition was enormous and unrelenting, yet it rarely seemed petty or self-serving. He genuinely believed in the importance of the human exploration of space, and saw our departure from Earth as part of the evolution of the human race. It is difficult to imagine how the Space Age would have unfolded in this century without von Braun. When he died it symbolized the passing of an era.

In the belt tightening that followed Apollo, the space station and any further human missions to the moon or Mars were all scrapped. NASA managed to save the shuttle, which, the agency argued, would make space travel more efficient and less expensive because it would be reusable and flown back to Earth by a pilot, features that led to its replacing all of the nation's disposable rockets—even the *Saturn V.* The irony was that during the 1970s when the shuttle was in development, human spaceflight halted in the United States. From the Apollo-Soyuz mission in July 1975 to the first flight of the *Columbia* space shuttle in 1981, no American flew in space.

Robotic missions, however, flourished. Thirteen Pioneer, Mariner, Viking, and Voyager probes departed Cape Canaveral between 1969 and 1978, partly as a result of the momentum created by Apollo. They orbited Mars and Venus, beamed back unprecedented views of Jupiter and Saturn, and even landed robots on Mars that searched, in vain, for life. Together they revolutionized scientific knowledge about every planet in the solar system except Pluto.

By the late 1970s, however, the shuttle project had begun to account for so much of NASA's budget that scientific missions began to suffer as the money supply shrank. This angered scien-

tists like James Van Allen and William Pickering who felt that the shuttle was squeezing science out in favor of more popular, but outrageously expensive, manned missions. To these men, and many others, the shuttle wasn't a spaceship, it was simply a high-tech white elephant that robbed science of the funds needed to continue the truly serious exploration of the solar system. Why build a shuttle, they wondered, when exploration could be accomplished so much more efficiently with smaller and simpler automated spacecraft launched by conventional rockets?

NASA, however, argued that the agency was more than an arm of the scientific community. Satisfying the human imperative to explore was also important, and a part of its original charter. Though the robotic probes represented fine and necessary achievements, the simple act of sending humans into space excited the public and generated support for the space program. In other words, without the drama of human spaceflight, money for pure science would be in far shorter supply. It was an old debate that dated back to the science vs. engineering arguments at White Sands, when the scientific purists were in one camp and von Braun was in the other.

Despite the scientific community's objections, the shuttle went into production. Officially it was referred to as the Space Transportation System (STS), as if it were a public transit operation, something ongoing and reliable. But from the start it was plagued by delays and cost overruns. The best and brightest scientists and engineers were assigned to the job, but the design and testing and construction became fragmented, divided among NASA's myriad centers and contractors, and it proved to be anything but cheap. Instead, some said the shuttle evolved into the world's most expensive hot rod, the Lamborghini of spaceships— a magnificent piece of engineering, but temperamental and astoundingly complex. Nevertheless it was expected to be absolutely reliable, and for five years and twenty-four flights it was. The American public took for granted that U.S. rockets no longer blew up as they had in the early days. Now there was a tradition of perfection.

Then on January 28, 1986, *Challenger* exploded. It happened ninety seconds into its flight, with the rocket roaring high into the sky over the Cape on an unusually cold Florida morning. A crowd of spectators and crewmembers' friends and family were watching. The shuttle pilot Commander Michael J. Smith had just said,

The space shuttle, officially referred to as the Space Transportation System, was designed to be recoverable, reusable, and cost-effective. The shuttle was an astounding engineering achievement, but the project was often plagued with delays and cost overruns.

"Go, you mother!" and then the ship detonated. Chunks of the fuselage hurtled into the Atlantic and tons of fuel vaporized in an orange ball. In the first few seconds no one knew, or wanted to know, what had happened.

On some level the shuttle program was designed to keep up the old momentum of Apollo, but the truth was that the U.S. manned space program had been stalled since Neil Armstrong planted the American flag on the moon in 1969. The glory days were gone, and the *Challenger* accident only seemed to underscore this disagreeable fact in the most dramatic and tragic way. The shuttle had proven neither inexpensive nor flexible; its purpose, therefore, became suspect. Now with *Challenger* it had also proven unsafe and unreliable. Seven astronauts had died, and there were serious questions as to why. Although the problems that caused the accident were ultimately discovered and repaired, although new and wonderful probes were eventually launched, another shuttle built, and the space station, the shuttle's intended goal, placed into development, *Challenger* really spelled the beginning of the end of the Space Transportation System. In 1991, the Bush administration announced that no more shuttles would be built, that the program was to be phased out over the next decade, and that NASA would return to a mixed fleet of disposable spacecraft.

Though it wasn't as spectacular as landing on the moon, the space shuttle has successfully set the concept of space exploration in the public mind as something that humans can do on a regular basis. The Soviet space station program, which evolved following the moon race and regularly ferried cosmonauts to the Salyut and Mir space stations, similarly helped make space seem a part of everyday life. As inconceivable as the idea of traveling and probing space had once seemed, it is now equally unthinkable to abandon it. It was during the post-Apollo years, when the shuttle was developed, that Europe, Japan, and China all began to organize aggressive, multibillion-dollar space programs that have now come to symbolize the status of a first-class nation, just as large armies and navies once did (though they are considerably less destructive).

The work during the shuttle years and the trauma of the *Challenger* accident also caused the United States and other nations to look beyond orbital launches and quick, flag-planting missions. Mars has now re-emerged as a serious destination for human exploration, and settlements on the moon and complex

Above: The space shuttle *Challenger* exploded on January 28, 1986, ninety seconds into flight. Seven astronauts died.

Opposite: Shuttle with open cargo bay. This picture was taken by a camera fixed to the helmet of astronaut Bruce McCandless as he coasted above the shuttle in his MMU (Manned Maneuvering Unit). He was about ninety meters away when the picture was taken.

Top: A transfer ship driven by a nuclear thermal rocket prepares to dock with a lunar lander. In the future similar vehicles capable of traveling to Mars could be built by expanding such a lunar ship's design, and then make their interplanetary journey in half the time a chemically powered rocket would.

Bottom: A nuclear-electric propulsion rocket the length of a football field fires its bank of ion thrusters as it enters Mars orbit. Powerful rockets like this one work by stripping atoms from a propellant, like Krypton, and then accelerating the ionized particles out of their thrusters. Ships like this do not exist, yet Robert Goddard pioneered the concept of ion propulsion early in this century.

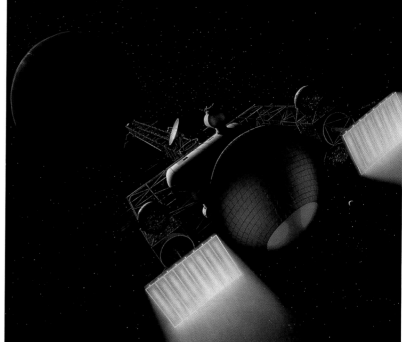

missions to the outer solar system are back on the drawing board. All of these ideas require investigating new breeds of spaceships that are more efficient, more powerful, and swifter than the chemical rockets that have done all of the work so far.

One alternative propulsion system is called a nuclear thermal rocket (NTR), a technology first tested in the 1960s by NASA and the Atomic Energy Commission (AEC), but a concept that traces its origins back to Robert Goddard, Konstantin Tsiolkovsky, and work done at Peenemünde. This rocket, developed by NASA and the AEC, was known as NERVA (Nuclear Engine for Rocket Vehicle Applications), and was powered by a nuclear reactor that generated enormous thrust. But testing on it stopped in 1973 when it became one more casualty of the tight budgets that followed Apollo.

Recently, however, the Department of Defense began to reinvestigate the nuclear rocket as part of its "Star Wars" research. An NTR can propel a spaceship twice as fast as the shuttle's chemical engines, and can cut its weight dramatically. This speed makes it enormously attractive for human missions to Mars, a factor that has also resurrected NASA's interest. High-speed ships would reduce a Mars-bound crew's exposure to lethal solar and cosmic radiation, lower the psychological stress of the journey, and lessen the time astronauts spend in zero gravity. Furthermore, a pound of nuclear fuel contains 10 million times more power than a pound of liquid hydrogen and oxygen, which vastly reduces the amount of fuel and supplies that would have to be hauled into space at great cost. (A portion of these reductions are offset by the increased weight of shielding needed to protect a crew from the radiation of a nuclear engine, but they still remain significant.)

The NTR works by pumping supercold liquid hydrogen ($-400°F$) into a chamber heated to $5,000°F$ by a nuclear reactor. Once there, the scorched hydrogen atoms instantly jump several thousand degrees, accelerate to supersonic speed, and explode out the engine's nozzle, creating enormous thrust. With this kind of power a Mars-bound crew could quickly reach a speed of 7 miles per second, and make the 35-million-mile trip to the red planet in 90 days rather than 180 or more.

Another new kind of propulsion currently enjoying a renewed interest is the ion engine, which could power ships designed to move large payloads over long distances. Ion engines trace their theoretical origins to Robert Goddard's secret paper,

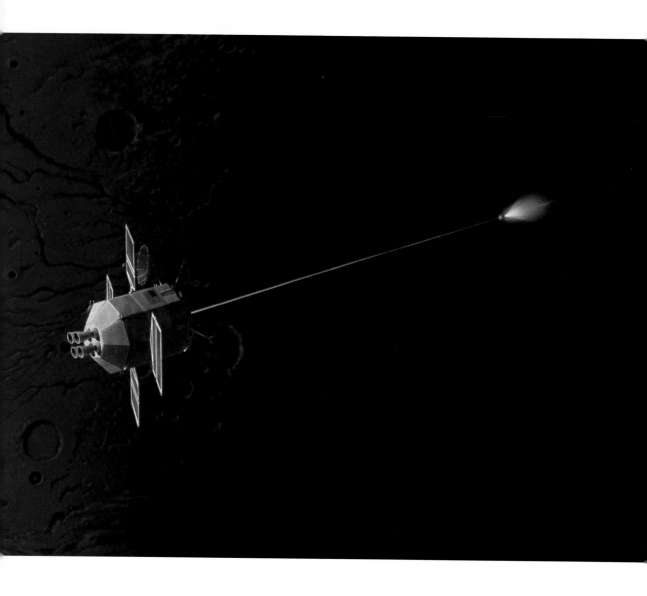

Three hundred miles above
Mars' Chryse Planitia (Plains of
Gold), a nuclear ship rendezvous
with a vehicle dispatched earlier
and now returning with samples
from another world.

Above: NERVA. The Nuclear Engine for Rocket Vehicle Application. During the sixties this nuclear engine was ground tested, but the project was abandoned in the early seventies.

Left: NERVA at its test site.

"The Ultimate Migration," but despite its impressive pedigree, the technology hasn't been perfected yet because it requires generating millions of watts of power at a steady rate over long periods of time, a feat extremely difficult to accomplish in space, where electrical power sources are limited.

Ion engines create thrust by feeding inert gases like krypton or argon into a chamber in which a tremendously powerful electromagnetic field strips one electron from each atom of the gas. This process, known as ionization, gives an atom an electric charge that allows a set of magnets in the thruster to accelerate them out the tail of the chamber. When they are ignited, ion engines don't have the tremendous kick nuclear thermal rockets do, but their acceleration is unrelenting and, if given enough time and room, can produce speeds of several hundred miles a second.

Not all futuristic spaceships rely on exotic hardware. One, in fact, harkens back to the sleek, wind-driven sailing ships of the last century. In 1924 Konstantin Tsiolkovsky first conceived of craft that could sail on impulses of light, but it was Carl Wiley, an aeronautical engineer at Rockwell, who independently spelled out the concept in a 1951 *Astounding Science Fiction* article entitled "Clipper Ships of Space."

Wiley described a spacecraft made of an extremely thin and reflective material unfurled in space that would catch light photons emitted by the sun. A photon's mass is minuscule; the force of sunlight on a solar sail the size of a football field, for example, amounts to the weight of a marble. But in space, where there is no drag and no gravity, that tiny force can produce remarkable acceleration. A square-rigged sail one-third of a mile on each side with the sun at its back would cover no better than 10 miles an hour in its first sixty minutes in space. But after a day it could travel at 200 miles an hour and after eighteen days it would be hurtling along at a mile a second and still picking up speed.

The first and most obvious beauty of solar sailing is that solar power costs nothing, and the hardware is simple—no engines or thrusters or pumps. The key technological problems center around designing a sail large enough and light enough to catch truly enormous quantities of sunlight. In 1976 NASA designed a lightship called the heliogyro, which it planned to rig with instruments to study Halley's comet on its swing near Earth in 1986, but the mission was scuttled for lack of funding.

Rob Staehle was a young engineer working at the Jet Propul-

sion Laboratory at the time, and found NASA's decision so disturbing that in 1979 he founded the World Space Foundation. Since the 1960s JPL had been in charge of designing all of NASA's unmanned interplanetary probes, but as far as he could see, NASA's sense of adventure had evaporated after Apollo, and he felt it was time someone put a little of the old-time swashbuckle back into space exploration.

He wanted to create an organization with a populist bent that could circumvent the red tape of government agencies and allow individuals to directly participate in a concrete effort to explore space. But Staehle found it difficult to finance the construction of spaceships by subscription, just as the VfR had, so he turned to solar sailing, which didn't require billion-dollar budgets.

Foundation members refer to solar sails as rebel technology because in time they could potentially make space exploration available to large numbers of ordinary people at reasonable cost, in the same way that covered wagons made the American West accessible to the millions of pioneers willing to ride them across the open prairie. The Foundation also feels that solar sailing technology could serve an important role in the future exploration of the solar system. For example, a sail one and a quarter miles square could conceivably haul the load of an eighteen-wheeler to Mars in four years and return empty twice as fast. Although such a feat wouldn't break any speed records, several of these sailing ships running like ferries between worlds could keep human outposts and expeditions fully supplied for a fraction of the cost of big chemical or nuclear ships.

To draw attention to the possibilities of their dream, the Foundation has recruited a corps of students, engineers, and consultants to design and build a working lightship that they hope will sail in what they call a sunjammer race to the moon scheduled to begin sometime in 1994. Organizations from Japan are also joining the race, and the hope is that a European Ariane rocket will deploy all the sailing ships in Earth orbit at the same time. The Foundation's ship is designed to unfurl a three thousand square meter golden mylar sail thinner than plastic sandwich wrap. Once deployed, the regatta's ships will begin to catch drafts of sunlight, swing around the Earth, and head off for the moon. Once at the moon (itself no mean feat) the Foundation hopes to guide its ship on to Mars, the first spacecraft not powered by a rocket to reach another planet.

A spinning solar sail known as the heliogyro was designed at the Jet Propulsion Laboratory in Pasadena, California, and was NASA's preferred solar sail candidate for a rendezvous to Halley's comet in 1986. Jettisoned from a space shuttle the craft would use twelve elongated sails which would ultimately extend four and a half miles from the spacecraft to form a giant solar pinwheel.

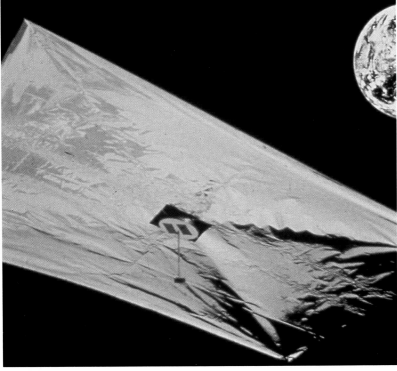

A solar sail could capture photons from the sun the way a canvas sail catches the wind. *Top:* A solar sail being built by the World Space Foundation for a proposed "sunjammer" race to the moon in 1994. *Bottom:* The ship as it might look unfurling its massive sail and beginning its journey from Earth to the moon. Theoretically such ships could achieve the speed of light.

Opposite: A solar sail ship on its way to Mars. Solar sailing ships could be used to haul cargo around the solar system.

Top left: The Tomahawk cruise missile in flight. The Tomahawk is the direct descendant of the V-2. It provides the U.S. Navy with a standoff, deep-strike capability against ships and land targets. It can travel in excess of seven hundred miles and carry a thousand-pound conventional warhead.

Bottom left: The Tomahawk missile over target.

Right: Tomahawk detonation over target.

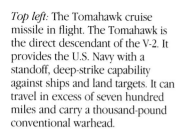

The hopes of the World Space Foundation harken back to the romantic impulses of the old rocket clubs and the dreams of the great pioneers. They have the lure of interplanetary adventure, and grassroots appeal. In the 1920s, the very people tinkering with rockets believed that they would soon be traveling in them. There was a certain naivete and purity in that kind of thinking. Seven decades ago when the first rocket clubs were forming, no one foresaw astronauts, or Vengeance Weapons, a cold war, intercontinental ballistic missiles, nuclear bombs, or even a space race. Events, however, conspired to join the lives of megalomaniacs with star-struck dreamers, politics with romance, destruction with deliverance. The exploration of the moon and planets was initially financed as a surrogate for war, a competition among superpowers, and having wished certain machines into existence, we became stuck with them, even in their most lethal forms.

On the other hand, there now seems to be a shift in the Space Age that is leading humanity in the direction of cooperation rather than cold-war-style competition. Exploring territory as expansive as the heavens may simply be a job that is too big for one nation, and many ventures in space, most notably the planned international space station Freedom, have become joint ventures. Perhaps in the future we will set out for other worlds less as Americans, Russians, Japanese, or Europeans and more, in the lingo of old science fiction novels, as Earthlings.

Biologists know that it is often in the nature of living things to cooperate, to grow increasingly interdependent—even unwittingly; each human being is, in fact, a monument to cellular symbiosis, 100 trillion cells joined in close partnership. Perhaps the Space Age itself is a part of a larger-scale version of this collaborative process. Or perhaps the picture is too large for us to ever know exactly where the urge to explore will eventually lead.

In any case there is hope. After all, despite the damage we have done to ourselves and our world over the past century, we have also managed an astounding bit of magic. We have transformed our primal curiosity about the stars into real ships that can ride beyond Earth and off to other worlds.

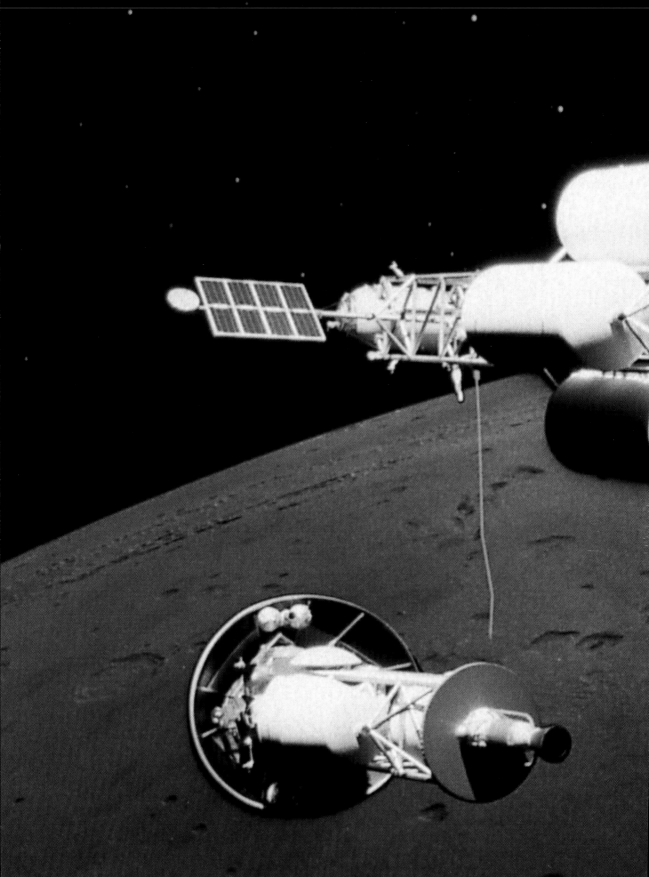

"It's part of the nature of man to start with romance and build to reality."

—Ray Bradbury

Quest for Planet Mars

Above: Martians, from a 1908 issue of *Collier's* magazine.

Left: Percival Lowell at his telescope. Lowell's childhood fascination with astronomy led him to build his own observatory in Flagstaff, Arizona, at that time one of the best in the world. He ordered its construction so that he could observe Mars as it passed within 35 million miles of Earth in 1894.

On the evening of August 22, 1924, William F. Friedman, chief of the code section of the Army Signal Corps, sat by his radio and listened intently for messages from Mars. Friedman was widely known for his ability to identify and decode radio signals—in fact, eighteen years later he would crack the Japanese diplomatic code just before the attack on Pearl Harbor. This particular night, however, he was the world's best hope for detecting alien transmissions.

To make his work easier, the United States government had ordered radio silence at all of its military installations around the world: from 12:00 A.M., August 22, to 12:00 A.M., August 25, the army and navy halted all radio messages just so that Friedman could pluck out a signal from across the cosmic abyss. But he heard nothing; not a crackle, or a hiss, not the slightest whisper of an interplanetary communication found its way through Earth's atmosphere to his expectant and talented ears.

Yet Friedman's assignment proved how strongly people believed, back in 1924, that intelligent beings lived on Mars. In the popular mind it wasn't so much a question of whether or not there were Martians, it was more a matter of what they were like and what they were up to. Only three years earlier, Guglielmo Marconi, acclaimed inventor of the radio, announced that he had received mysterious signals—some of them in Morse code—from a source he couldn't identify. The public naturally assumed that since there weren't many earthly sources of radio transmissions in 1921, Martians must have somehow picked up Earth's Morse code communications and were now trying to tap out a response to their interplanetary brethren.

But if Friedman's efforts in 1924 were any indication, this assumption was wrong. Mars's inhabitants were either extinct or disinterested. One astronomer, a flamboyant Frenchman by the name of Camille Flammarion who believed fervently in a race of wise and ancient Martians, concluded, "Perhaps the Martians tried to communicate with Earth before, in the epoch of the iguanodon and dinosaur, and got tired."

We on Earth have a long-standing love affair with Mars. For more than a century we have been enthralled with the idea that it harbors life. In our collective consciousness Mars is not simply a planet, an astronomical curiosity, but a mystery and a beacon, a place as myth laden as Olympus or Valhalla or Shangri-La. We still feel this way today, largely because of a turn-of-the-century aristocrat named Percival Lowell, a man smitten with by a passion for astronomy and gripped by an interplanetary sense of drama.

There was nothing about Lowell that would have led to the conclusion that he would become one of the great purveyors of complete hokum in this century. In fact, he was the epitome of mainline New England respectability, a Brahmin whose parents were the central pillars of Boston society. The first Percival Lowell had sailed to North America with the initial wave of Puritan settlers back when Galileo was still alive, and by the early nineteenth century his family had amassed one of the great fortunes in the country by founding America's textile industry. By the time the second Percival came along in 1855, the name Lowell and the word *millionaire* were synonymous. Soon he would make the name synonymous with Mars too.

Lowell looked every inch the aristocrat, impeccably dressed with intelligent, confident eyes, bald pate, and a great dense mustache that swept away from his patrician features. He graduated from Harvard Phi Beta Kappa, his mathematics teacher called him the most gifted student he'd ever taught, and his cousin, James Russell, the editor of the *Atlantic Monthly* and a respected author, called Percival "the most brilliant man in Boston." He was a blue-blood through and through.

Even as a child Lowell had been fascinated with astronomy. In fact, he liked to tell people that his earliest memory was of Donati's comet, which, with its great scimitar tail, was one of the great celestial light shows of the nineteenth century, streaking

Top: Giovanni Schiaparelli's 1877 map of the planet Mars. Schiaparelli saw markings on Mars that he called *canali*. In Italian this means natural channels of water, but in translation it became canals, suggesting that the planet was inhabited by intelligent life.

Bottom: Lowell's map of Mars reveals the canals that he felt had been built by a desperate and heroic race of Martians to delay their own inevitable extinction. Lowell believed water from the polar caps was pumped to the desert regions of the planet.

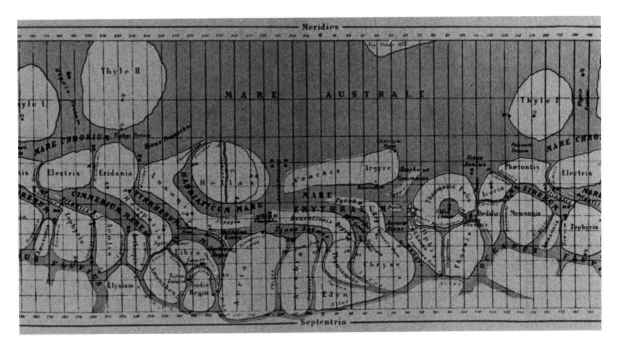

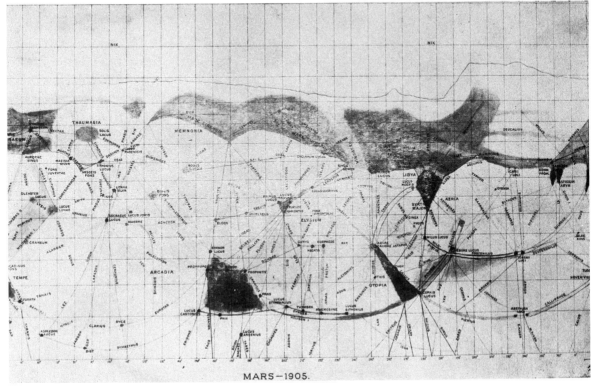

MARS—1905.

Above: Camille Flammarion. In 1892 Flammarion published a review of all observations of Mars entitled *Le Planete Mars*. He wrote: "It is the first time since the origin of mankind that we have discovered in the heavens a new world sufficiently like the Earth to awaken our sympathies."

Opposite: A globe of Mars created by Lowell. "Its smaller bulk has caused it to age quicker than our Earth," he wrote, "and in consequence, it has long since passed through that stage of its planetary career which the Earth at present is experiencing, and has advanced to a further one to which in time the Earth must come."

across the night sky for the better part of four months in 1858. This fascination ultimately inspired Lowell to tap his personal fortune in 1893 and build his very own observatory in Flagstaff, Arizona. After years as a traveler, diplomat, and successful author, this observatory became his way of resolving an ongoing debate among scientists about the possibility of life on Mars. Lowell recruited some of the top talent in the field, bought a plot of land on a mesa seven thousand feet high, and rushed an eighteen-inch refractor telescope and a state-of-the-art observatory into construction. The facility had to be ready by 1894, in time to catch sight of Mars at its closest approach to the Earth, when it is only 35 million miles away.[5]

Up to this point, Lowell had gathered most of his information about the red planet from the writings of the distinguished Italian astronomer Giovanni Schiaparelli who described strange features he had observed on Mars fifteen years earlier. Schiaparelli called these odd linear configurations *canali,* which means grooves or channels in Italian (as in the English Channel), and described a world in which masses of land lay separated by water, not unlike Earth.

Lowell, however, misinterpreted Schiaparelli's *canali* to mean canals—man-made waterways. In the late nineteenth century, you could hardly have chosen a better symbol than a canal to represent an intelligent being's ascendancy over nature; they were the hallmark of advanced civilization. The Suez Canal was completed in 1869, the Corinth Canal was completed in 1893, and plans were in the works to undertake the greatest engineering feat of all—the construction of the Panama Canal, which would link the Atlantic with the Pacific.

After many long nights squinting through the long barrel of his telescope on the crest of what he had christened Mars Hill, Lowell began to make out a network of waterways on the Martian surface, and that led him to imagine that a terrible tragedy was unfolding there.

Lowell's calculations showed that Mars's mass was much less than Earth's, which, he theorized, meant that the planet was rapidly losing water vapor. Its oceans were literally disappearing into thin air, a phenomenon that he believed had accelerated Martian

[5]   Because Earth and Mars travel at different speeds around the sun (Mars takes almost twice as long to make one revolution), the distance between the two planets is always changing. The closest they ever come to each other is 35 million miles when Mars lies in alignment just beyond Earth; the farthest is 249 million when the two planets lie at opposite sides of the sun.

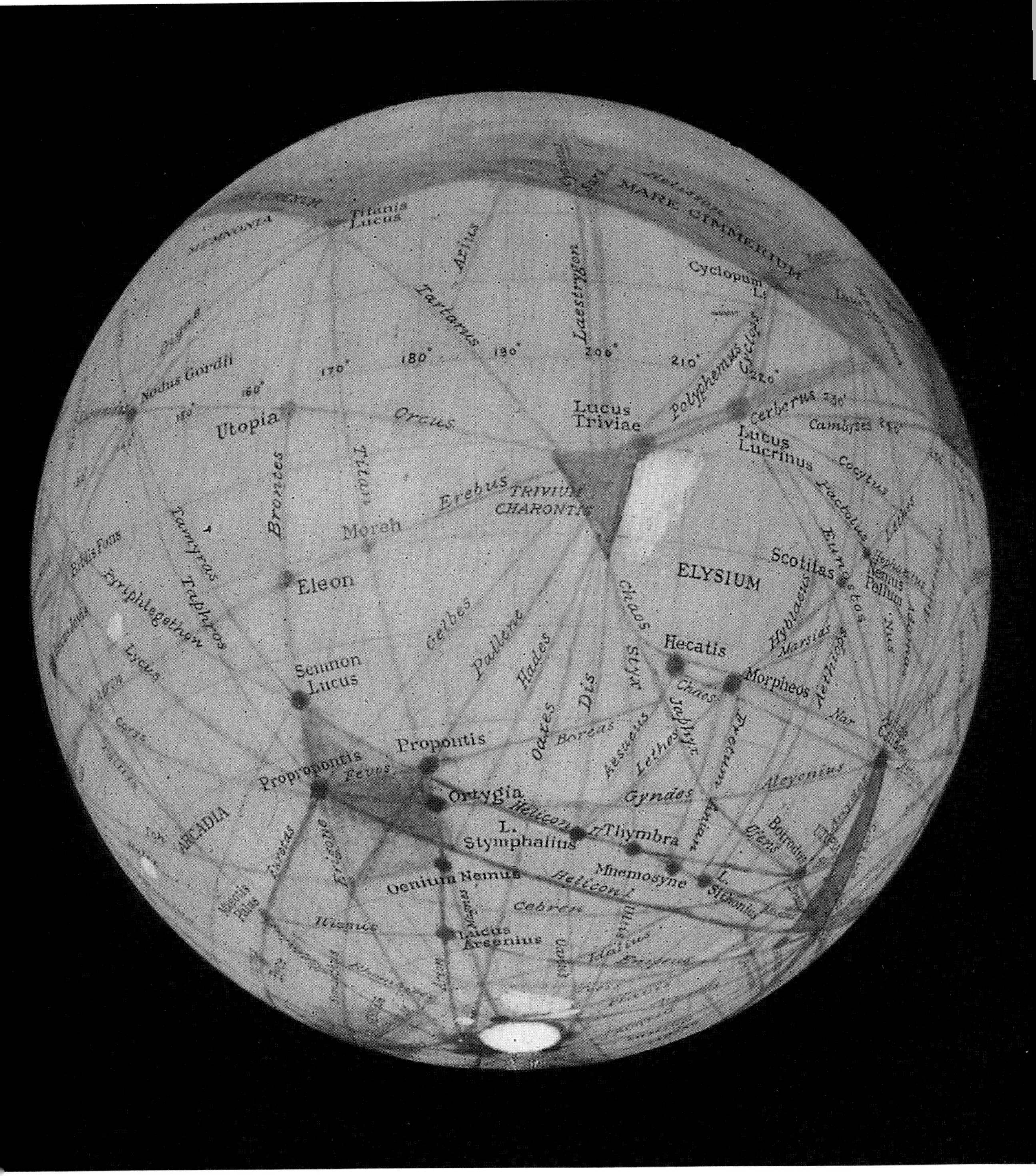

MARE CIMMERIUM
MEMNONIA
Titanis Lucus
Arius
Tartarus
Laestrygon
Cyclopum
Polyphemus
Cyclops
Nodus Gordii
160°
170°
180°
190°
200°
210°
220°
230°
150°
Utopia
Orcus
Lucus Triviae
Cerberus
Cambyses
Lucus Lucrinus
Cocytus
Titan
Brontes
Erebus
TRIVIUM CHARONTIS
Moreh
Tamyras
Taphros
Pyriphlegethon
Eleon
ELYSIUM
Pactolus
Eunostos
Scotitas
Nemus Pelium
Lethes
Chaos
Hyblaeus
Marsias
Aethiops
Gelbes
Pallene
Hades
Dis
Styx
Hecatis
Morpheos
Sennon Lucus
Lycus
Corys
Oaxes
Boreas
Aesacus
Lethes
Propontis
Protonilus
Nar
Propropontis
Fevos
Ortygia
Helicon
Thymbra
Gyndes
Alcyonius
ARCADIA
L. Stymphalius
Mnemosyne
Botrodus
Ufens
Euratas
Erigone
Oenium Nemus
Helicon I
L. Sithonius
Cebren
Hissus
Lucus Arsenius
Idalius
Enipeus

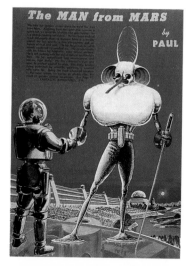

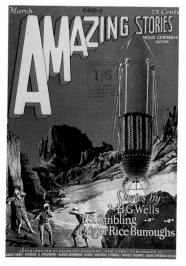

Top: A Martian greets an alien from Earth . . . *Bottom:* and an interplanetary spaceship takes off. Lowell's Mars sparked a torrent of science fiction in the 1920s and 1930s in magazines like Hugo Gernsback's *Amazing Stories* and *Science Wonder Stories.* Magazines like these proliferated, supported by a readership fascinated with the possibilities of extraterrestrial life. Mars became a place where anything was possible.

evolution, and therefore brought forth a race of highly intelligent beings. As the planet's vast seas evaporated into little more than plateaus of sand and dirt, the only sources of water left on Mars lay at its north and south poles. Through his telescope Lowell had seen snowy glaciers at both ends of the planet, but had also noticed that during the Martian summer a deep blue crescent formed along the southern ice cap. This, he believed, was an enormous pool of melted snow. As the summer progressed the ice cap disappeared altogether, which to Lowell could only mean that as the snow melted it was being siphoned off and pumped across the planet, through a labyrinthine plumbing system, to the equator and the Martians' few surviving cities. He concluded that an engineering feat of this magnitude would require such thorough global cooperation and so much technical know-how that it must mean that Mars's inhabitants were a wise and peaceable race, desperately working to slow what they undoubtedly knew, in their advanced state of intelligence, was the inevitable end of their kind.

When Lowell revealed his findings to the world in his first book about the red planet, entitled simply *Mars,* the publisher couldn't get enough of them into print. The unequivocal statement that there was intelligent life elsewhere in the solar system was astounding enough, but his description of a dying race had a mythic irresistibility about it which left readers spellbound. Amateur astronomy exploded and telescope sales skyrocketed. People all over the world sat in their backyards squinting behind their two-and-a-quarter-inch spy glasses, playing out in their minds this alien, interplanetary soap opera. Strange events followed. One reader of the New York *Tribune* claimed to have deciphered the ominous words *the Almighty* etched in Hebrew out of dark markings on Mars's surface. . . . Were the Martians calling to God? (And how had they come to know Hebrew?) About the same time certain astronomers detected light signals from the parched Martians, interplanetary SOSs beamed, the theory went, from a marble obelisk several miles high.

Wild speculation erupted on ways to establish contact. One suggestion called for inscribing the Pythagorean theorem at half the size of Europe across the Sahara desert to let the Martians know that Earthlings were willing to help them with their plight. How any subsequent acts of interplanetary charity would actually be achieved, however, wasn't addressed.

Critics of Lowell's theories pointed out that not even an instrument as powerful as an eighteen-inch refractor telescope could reveal thin canals of water on Mars. Lowell agreed, but explained that he wasn't seeing the canals themselves, only the bands of vegetation that grew alongside the canals, the "denser verdure athwart them." The skeptics were shocked that people believed Lowell at all. E. E. Barnard, another respected astronomer at the time, wrote that he didn't see any canals, only a few dark features and elevated plateaus. How could Lowell and Schiaparelli and Lowell's staff keep seeing these rigid lines in the same places year after year, when he knew they simply weren't there?

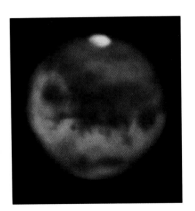

The scientific community by and large lined up behind Barnard, or at least they strongly questioned Lowell's great melodrama. In fact, based on writings before his first observations, it looked suspiciously as if Lowell saw precisely what he wanted to see regardless of the evidence. Still, despite his penchant for drama, there were some speculations he refused to make. He did not pretend to know anything about the physical characteristics of Martians, for example—that would have been going too far. "For answers to such problems," he wrote, "we must look to the future."

Others, however, willingly speculated, or, more accurately, wildly fantasized about such questions, and more. The king of these fantasies was a man named Edgar Rice Burroughs, a novelist best remembered today as the author of the Tarzan books, but a writer who initially made his reputation with a series of adventure novels that unfolded on the richly alien world of Barsoom. Barsoom, in Burroughs's universe, was what the Martians themselves called their planet.

In 1911, at the height of Lowell's speculations on Mars, Burroughs began writing these grand pulp operas, and the public swallowed them whole. These novels featured a gentleman adventurer named John Carter who, while facing certain death at the hands of Apache warriors in the Arizona desert, simply imagined himself across space to the desiccated surface of the red planet. On a world where no human had ever set foot, Burroughs's hero wandered the spired and ancient cities of Barsoom, navigated its life-giving canals, and engaged in combat with everything from fifteen-foot warriors called Tharks to four-armed white gorillas singularly intent on his demise. Would he save the stunning and buxom Dejah Thoris, Princess of Helium? Could he overcome the dark and alien forces beneath "the hurtling Moons of Barsoom"?

The best photograph of Mars taken from an Earth-based observatory.

Readers never knew, until the books' final pages.

The Barsoom novels were pure fantasy, but they brought Lowell's Mars alive, and won the hearts and minds of a generation of eight-year-olds, some of whom, like astronomer Carl Sagan and novelist Ray Bradbury, found enough inspiration in them for whole careers. As Bradbury put it, John Carter "seized off to Mars by impossible dreams, . . . and changed America's scientific territory forever." Even after Lowell died in 1916 and his ideas had fallen out of scientific favor, Burroughs continued to nail Mars firmly into the public consciousness. The popularity of his books helped set the entire modern science fiction movement into motion, including magazines such as Hugo Gernsback's *Amazing Stories* and *Science Wonder Stories,* which enjoyed remarkable popularity thanks to a readership fascinated with the pregnant possibilities of Mars. These in turn begot a new generation of virtuoso authors such as Robert Heinlein, Isaac Asimov, Ray Bradbury, and Arthur C. Clarke who went on to cut their imaginations loose on the infinite terrain of the universe: every one of them wrote about Mars, and with every story they wove it more deeply into the cultural fabric.

Few places, extraterrestrial or otherwise, have claimed so much of our imaginary territory or sparked more scientific speculation. The science fiction stories and the films that followed Lowell, particularly the best of them, made Mars a place where anything was possible, a locus of magic where we could play out our worst fears, greatest hopes, and most profound questions. Maybe that is why it seems inconceivable that we could ever forgo its exploration. We sense that in exploring Mars we may uncover more secrets about ourselves than anything else, and that it will lead, like the old myths themselves, to a better understanding of where we fit in the scheme of things.

If you had asked a leading planetary scientist in 1960 to explain what was known about Mars, he or she would have said something like this: It is pretty flat, with a few rolling hills at most. The atmosphere is composed mainly of nitrogen and exerts a pressure similar to Earth's. The polar caps are nothing more than a wafer-thin hoarfrost. The dark areas are old sea beds probably filled with primitive organic material—vegetation of some kind—and there are naturally formed channels of water.

Every one of these beliefs was wrong—that's what happens

The first of twenty-two images of the Martian surface returned by *Mariner 4* eight months after its launch on March 28, 1964. They provided our first close-up look at another planet.

when you look at something from 35 million miles away.

The real truth about Mars began to emerge in 1965 when a U.S. probe called *Mariner 4* raced within 6,118 miles of the planet, providing the first good view Earthlings had ever had of the planet. Not that we hadn't been trying. From the moment *Sputnik I* had been rocketed into orbit eight years earlier, the exploration of Mars had been at the top of the space agendas of both the Soviet Union and the United States. But successes were hard to come by. After several failed attempts by the U.S.S.R. and one by the United States, scientists began to joke that there was some force out there —they called it the Ghoul—a monster determined to thwart all of their efforts. Between 1960 and 1965 no less than six probes failed to reach their objective and information on Mars remained Earth based and inaccurate.

Then in 1965 NASA's *Mariner 4* finally reached the red planet and beamed back twenty-two images that revealed no sign of vegetation or rivers, and no canals—just a frozen world with a pockmarked surface that apparently hadn't changed since it had been hammered with incoming meteors 4 billion years earlier. The impression the images gave was that the main difference between the moon and Mars was that Mars was red.

Mariner 6 and *7* soon followed and confirmed all of these findings, crushing the hopes of scientists who had still hoped that some simple life-forms might have managed to eke out an existence on the battered planet. But all the evidence indicated that the planet was ice cold and hadn't changed since the beginning of time. Then came *Mariner 9.*

When *Mariner 9* reached Mars on November 13, 1971, the planet was engulfed in a furious global dust storm that rendered the surface invisible. Winds were clocked at over 300 miles an hour, and there wasn't much the *Mariner* teams could do but wait by their monitors at the Jet Propulsion Laboratory as the probe beamed back image after image of an erased world.

At about the same time *Mariner* went up, the U.S.S.R. had launched two probes, *Mars 2* and *Mars 3,* which were preprogrammed to land on the planet. When the robots arrived their trajectory didn't allow them to check Earth for further instructions, so they dutifully plummeted through the windstorm and disappeared. *Mars 3* actually managed to reach the surface, making it the first Earthly object to land on Mars, but it only coughed back twenty seconds of information before it, rather literally, bit the dust.

Top: Olympus Mons, the largest of the Martian volcanoes, peeks through an alien cloud deck. Olympus Mons measures 435 miles across its base and rises fifteen and a half miles into the sky, nearly three times the height of Mt. Everest.

Bottom: Valles Marineris, named for the *Mariner 9* spacecraft that discovered it. The valley is more than three thousand miles long and up to 435 miles wide, ten times the size of the Grand Canyon. Some sections of this canyon are five and a half miles deep.

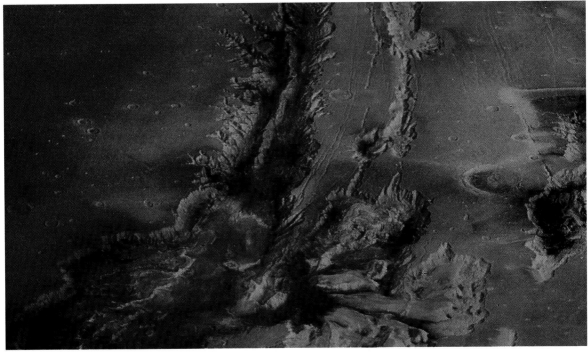

Mariner 9, which was never designed to land on Mars, simply parked itself in orbit and circled for three months while the storm raged on. The whirling clouds of red dust that enveloped the planet were frustrating, but the violent weather itself provided the first indication that Mars was more than a ruddy version of the moon.

Slowly, grudgingly, the storm relented, and as it did four strange protuberances began to appear through the whirling haze, the tops, it turned out, of tremendous volcanoes made up of colossal mounds of lava. The summit of the largest mountain, Olympus Mons, rose fifteen miles into the sky—three times higher than Mt. Everest—and its base was as wide as Montana. Southeast near the Martian equator scientists also found a gargantuan canyon that cut a jagged gash in the planet twenty-five hundred miles long and, in some places, five miles deep, a feature that could easily swallow the entire Grand Canyon in one of its larger tributaries. Scientists dubbed the scar the Valles Marineris for the probe that discovered it.

Mars, however, saved its best secrets for last. As the dust settled, *Mariner* revealed hundreds of miles of channels and grooves east and north of the Valles Marineris that looked like they had been formed by running water. Some were shaped like dried riverbeds on Earth with treelike branches, but many seemed to have sprung out of nowhere, their channels equally wide from beginning to end. Both types stunned scientists. Features like these made no sense on a world as dry and cold as Mars where running water would be impossible. So how did they get there?

Strangely enough, one answer came from the utterly unrelated work of a tenacious old geologist named Harlen Bretz. Beginning in the 1920s Bretz had spent much of his career studying a bizarre thirty-seven-hundred-square-mile area in Washington State that he came to call the Channeled Scablands, a place strewn with house-sized boulders and scarred with enormous ravines, towering, jagged cataracts, and great rippling ridges of sand and gravel.

Bretz was determined to find out how the landscape had been so scarred, and after years of study, concluded that thirteen thousand years earlier, during the last ice age, a lake had formed along the Montana-Idaho border behind a glacial dam, and one day the dam burst. Three thousand square miles of water, three hundred feet high, suddenly exploded across the Idaho panhan-

Top: Landsat photographs of the southern Scablands, which cover roughly twelve thousand miles of Washington State. The terrain was created by a catastrophic flood when a glacial lake in Montana burst its ice dam thirteen thousand years ago and sent a wall of water three hundred feet high exploding across the Idaho panhandle and the state of Washington into the Pacific Ocean.

Middle: A view of some of the Scablands channels created by the ancient flood. Today rivers flow through sections of the scarred landscape, but it wasn't until the 1970s that scientists understood what created them.

Bottom: Viking mosaic of channels on Mars probably formed by floods like the one that created the Scablands. If Mars once had water, did it also harbor life?

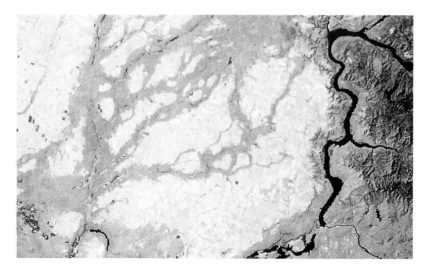

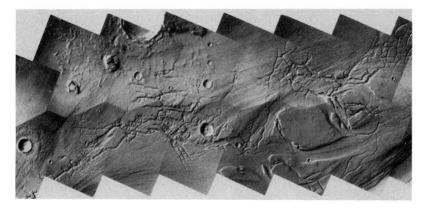

dle and through Washington State into the Pacific, leaving the ragged landscape behind it.

It took decades, but Bretz finally managed to win acceptance for his theory and shortly thereafter planetary scientists began to compare photos taken of the scablands by a sophisticated earth-orbiting satellite called Landsat with *Mariner*'s images of the enormous Martian channels. They looked so similar that some scientists began to suggest that leviathan floods had also formed the strange features on Mars. No one could say what had initiated the Martian deluges in the first place, but based on the evidence, some theorized they were created when meteors plummeted into the planet or volcanoes suddenly erupted, transforming the Martian permafrost into surging oceans of water and mud.

The catastrophic-flood theory explained the wide, gouged channels that seem to have sprung out of nowhere, but not the more ancient, Earthlike riverbeds that *Mariner*'s cameras had found. These, most scientists now believe, could only have been formed by running water that existed when the planet was warmer. But if it were the case that Mars once had rivers of water, couldn't life have also established a toehold there as it had on Earth? Two missions would soon be mounted to find out.

Viking 1 and *2* were highly sophisticated probes designed as a kind of space age, cybernetic tag team. Both would circle the planet then jettison landers, which would descend to Mars to perform a number of tests including the most anticipated of all: make contact.

The irony of dispatching nonliving machines built by living beings on one world to search out life-forms on another wasn't lost on everyone. But in the post-Apollo era, NASA's funds for human missions to Mars had been heavily curtailed. By the mid-1970s, the scientific community was elated simply to have preserved any congressional money at all for interplanetary missions.

Though the Vikings weren't human they were true marvels of science and engineering, so adaptable that their final landing locations were chosen only hours before they descended, based on information the orbiters themselves had observed and provided. No Earthly object in the long history of the human race had ever successfully landed on the surface of another planet, and the pictures the Vikings beamed back—slowly, line by line—absolutely stunned the scientists looking on. It was one thing for these

Top: The first photograph ever taken on the surface of Mars by *Viking I* minutes after it landed July 20, 1976. That is Viking's footpad at the lower right.

Bottom: Frost on the Martian landscape. Scientists believe dust particles in the atmosphere pick up bits of ice. Carbon dioxide, which makes up 95 percent of the Martian atmosphere, then freezes and adheres to these particles making them heavy enough to sink to the ground. Warmed by the sun, the carbon dioxide evaporates and returns to the atmosphere leaving behind the ice and dust.

men and women to view satellite photographs taken hundreds of miles above the planet, images that revealed Mars from a grand and global scale, but it was another matter to look upon the landscape of this world, so deeply rooted in their collective psyche, from a human perspective, as though they were standing right there on the dust-blown surface itself.

Somehow these pictures looked familiar. Not at all unlike the Mojave desert with its broad, rocky, and reddish landscape. The probes revealed no yellow vegetation as Burroughs had envisaged, no princesses in distress, no Tharks, but it was exciting nevertheless. Astronomer Carl Sagan later recalled, he might not have been surprised to see a prospector with his mule in tow amble across the horizon. Not that that would have been possible. It would take a hardy creature to survive in these plains where in the blazing afternoon sun, at the height of summer, the temperature barely crawled up to $-24°F$. Furthermore, 95 percent of the Martian air consisted of carbon dioxide, and only about one-tenth of one percent was oxygen. If a human stepped out onto the landscape unprotected, his last lungful of air would be instantly sucked from his body, and he'd be dead in thirty seconds. Looks can be deceiving.

Still, as some scientists looked on, they wondered if some dauntless microbe might have outmaneuvered the harsh course of Mars's cosmic evolution and managed to scratch out a living beneath the soil of this noxious, planetary freezer. On Earth a handful of ordinary dirt from an average backyard has more microbes in it than the Milky Way has stars. Could some uni-celled creature have survived in the red regolith of Mars? The Vikings were designed to find out. Each probe had been rigged with a miniature laboratory and a robot "hand" that was to scoop up a sample of the alien planet for testing. This, scientists hoped, would resolve the issue.

Once they arrived both probes went to work like cool and resolute gem cutters on the blasted, barren Martian landscape. They scooped up their samples, and their tiny internal laboratories tested them for evidence of life, laboring to negotiate a meeting between Earthlings and Martians. They each performed three experiments. The first soaked Martian soil in a solution of vitamins, amino acids, purines, and organic acids to see if any microbes would eat the nutrients and release gases just as a plant exhales oxygen or a human exhales carbon dioxide. Amazingly, the tested sample did in fact exhale large amounts of oxygen. The next day a second experiment mixed similar nutrients with the Martian soil, except these nutrients included some radioactive carbon atoms. If Martian microbes ate any of the radioactively tagged carbon, the probe's instruments would detect them in the radioactivity that the microbes released. Soon after the nutrients were mixed with the soil the chamber billowed with radioactive gases, which again indicated that something must be digesting the mixture. As an alternative to the earlier experiments, which could have drowned potential Martian microbes, the final experiment simply cooked the soil in an atmosphere that was enriched with radioactive carbon dioxide. If the microbes breathed the radioactive atmosphere, assimilating it somehow like a plant assimilates carbon dioxide, it might show that something was alive. Viking cooked the soil for five days and, in the end, detected a small but measurable increase in radiative gases.

All of these tests seemed to say that, at long last, life had been discovered on Mars—except that another instrument on both landers contradicted these findings. A gas chromatograph–mass spectrometer had already told scientists there wasn't a single molecule of organic material anywhere within reach of either of the Viking probes (which had landed thousands of miles apart), and where there were no organic molecules, there was very little chance of finding life, at least as we know it.

This was perplexing because, frankly, it was difficult to comprehend how Mars could *not* have any organic material on it. Organic compounds are essentially carbon compounds; in fact, they are called organic because all life on Earth is carbon based. Mars orbits on the inner edge of the solar system's asteroid belt, which is nothing but shattered cosmic debris made up largely of carbon. It was standard theory that Mars had been heavily bombarded by this debris over billions of years just as Earth had been, and that the carbon from this bombardment was central to the genesis of life. But if there was no organic material on Mars, either from orbital debris or living creatures, how could all three biological experiments have turned up positive results?

The Viking scientific teams puzzled over this conundrum. Finally, they reckoned that in the case of the first two experiments the discrepancy was caused by some anomaly in the Martian chemistry. The soil on Mars was cold, dry, and heavily bombarded by solar radiation, factors that could have produced unpredictable results when the soil and soups were mixed.

Explaining the outcome of the third experiment was more difficult. Even after the soil was heated far beyond a temperature that would have annihilated any living thing on Earth, the radio-active gases still wafted up from the soil. It didn't make sense that anything could have survived the heat, but there wasn't any ready alternative—maybe some hardy microbe lay burrowed in the soil.

After analyzing and reanalyzing, the Viking biology team could only report that the results simply didn't "permit any final conclusion about the presence of life on Mars." Despite our advanced technology and all of the intricate cybernetics, Mars had once again confounded its terrestrial inquisitors. The twin Viking probes had covered 35 million miles, sought their answers, and, where the mysteries of life were concerned, had only come up with more questions. Since then no other landers have returned to Mars to press the point, and the question remains unresolved. Scientists *have,* however, uncovered some interesting clues from a second unexpected source—the Earth.

INVASION OF THE INSECT ROBOTS

Since the days of the Viking landers, which were the last space probes to visit Mars and successfully beam back information from its surface, countless designs for the next machines to explore the red planet's terrain have been presented to NASA, everything from contrivances that hop to devices that crawl and roll, look like enormous axles with gargantuan wheels, or resemble giant grasshoppers. Whatever their final configuration, robots of some kind will probably visit Mars before humans do, and at least one of them will be designed to gather samples of rock and return them for inspection on Earth.

Among the robots that have been tested for this job is a six-legged machine at Carnegie Mellon University in Pittsburgh called Ambler that can walk, gingerly, through very rough terrain by circulating two sets of three metal feet. NASA has also contracted the Jet Propulsion Laboratory to build a more conventional rover that would be outfitted with wire mesh wheels to withstand Mars's numbing cold (rubber would shatter). Both machines are designed to return samples of

Fossil traces in the Dry Valleys in Antarctica show that the valleys were once a very different place. Scientists previously believed that nothing could survive here, but since the early 1970s they have found several varieties of microbial life, all that remains of the richer environment that once existed.

Martian rock, which means they would have to be outfitted with sophisticated stereoscopic cameras that can register depth, radar sensors that can determine the difference between solid ground and a sand pit, and a silicon brain loaded with a complex map of the area it is exploring. The map would be supplied by orbital satellites that would be sent ahead of the rover, because robots on Mars will not be able to be operated from Earth by remote control: It takes at least twenty minutes for radio signals to travel between the two planets, a time delay that could result in the robot careening to the bottom of a chasm before its operator on Earth had even received the machine's last transmission. Because of their complexity and size, scientists have their doubts about the likelihood of these particular machines someday roaming Mars, but progress is being made.

A radically different, but promising method for exploring Mars robotically has been worked out by a maverick engineer at MIT named Rodney Brooks who is building not large, complex robots but small buglike ones that are relatively dumb. Unlike most experts in the robotics field, Brooks doesn't believe it is important to attempt to mimic real human intelligence. Instead he concentrates on building foot-long, six-legged robots that look like futuristic insects and operate according to very simple rules. These machines never consult a plan or map, but simply move by following basic knee-jerk reactions programmed into them; commands like MOVE FORWARD, BACK OFF, LIFT (leg) HIGHER, ROTATE JOINTS.

"Instead of building the ultimate vision system and the ultimate planning system and the ultimate execution monitor, I decided to build a robot that could go down a corridor without hitting stuff, even when people were getting in its way."

Brooks would like to dispatch a corps of these small robots to the moon and Mars with cameras attached because they are cheap, would spread the risk of failure among many machines, and with the benefit of a map downloaded from a satellite would operate independently. NASA hasn't altogether accepted Brooks's concept yet, and perhaps never will, but the image of landing a spaceship on Mars and having a corps of insectlike creatures emerge from it seems almost too perfect to pass up.

Top: Microbial mats on the bottom of a perennially ice-covered Antarctic lake. Strange and unexpected life-forms.

Bottom: Ice-capped Lake Vanda in the Asgard Range, Victoria Land, Antarctica, not far from Lake Hoare in the Dry Valleys. Research scientists have found hardy life-forms at the bottoms of these lakes. Maybe as Mars froze, a few bodies of water and colonies of tenacious Martian microbes also managed to survive under caps of ice.

Nearly everyone pictures Antarctica as an enormous dome of ice at the bottom of the planet, and most of the continent fits that description. But large sections are also mountainous, and huddled among their black peaks are areas called the "dry valleys," bitter cold deserts, majestic in scale, but absolutely bleak—the driest and coldest places on Earth. Some of the valleys haven't seen rain or snow for two million years because the mountains that surround them strip the moisture out of the winds as they whip in from the coast. Winter temperatures drop to $-112°F$, which makes it a deadly place. In fact, for decades most scientists considered the valleys utterly sterile. But in the early 1970s they were proved wrong when an astoundingly tenacious form of yeast was found living beneath the surface of certain rocks in the valley. Somehow these invisible creatures managed to draw just enough water from the parched air, glean just enough nutrients from the rock, and gather in just enough warmth from the stingy Antarctic sun to colonize the stone.

This discovery was made by Imre Friedman, a biologist from Florida State University, who together with a colleague named Chris McKay, a planetary scientist at NASA's Ames Research Center near San Francisco, began to theorize that if these tough microbes could find a way to survive in Antarctica, perhaps life on Mars had somehow done the same.

Then a few years later another form of life turned up, this time in Lake Hoare, one of the dry valleys' many frozen lakes, which were created over thousands of years by the snow melt that trickles down from the surrounding Transantarctic Mountains. Even in the warmest part of the year the lake is capped with a lid of ice ten feet thick, but this cap allows the summer sunlight to prevent the lake from freezing completely even during the sunless Antarctic winters. There was no reason to believe that anything could possibly live in a place this remote and frigid, but in 1978 biologists found the bottom of the lake covered with enormous mats of microbial communities resembling the colonies of primitive organisms called stromatolites that were among the first and most dominant forms of life on Earth three and a half billion years ago.

Given half a chance, it seems that life will establish itself anywhere it can, no matter how hostile the surroundings. In fact, it turns out that other frozen lakes throughout the valleys are also alive, sustaining a bestiary of algae, bacteria, and fungi by capping

Chris McKay of the NASA Ames Research Center peers down into the ten-foot-deep hole in the ice of Lake Hoare. A three-foot-wide metal coil filled with hot antifreeze created the tunnel that opened up the lake.

the sun's heat, concentrating dissolved gases, and, in some cases, raising their water temperature as high as 77°F. It occurred to Friedman and McKay that if bodies of water like these still exist in the dry valleys, maybe similar lakes once also existed on Mars, a possibility that inspired them, and others, to take a harder look at the photos returned by the Viking probes. Sure enough, their investigations turned up places in the Valles Marineris that look like dried lake beds, desiccated Martian versions of Lake Hoare. Investigating them would require very sophisticated explorers, according to McKay, work well beyond the abilities of even the most complex robots. A search for life in these places would mean sending to Mars not machines, but human beings.

After the Viking missions, in the 70s, talk about sending humans to Mars continued, but talk was nothing new. Scenarios of this sort have a long tradition, dating back to Wernher von Braun's legendary 1952 paper "The Mars Project," the first realistic proposal for mounting a human expedition to the red planet. Countless other drawing-board journeys followed, and by now all of the scientific and government reports that have outlined how, when, and why human missions to Mars should take place could fill a good-sized bank vault.

NASA's very first long-range plan in 1959 nimbly skirted the actual mention of a manned mission to Mars. It simply assigned the idea to nameless futures that would emerge after 1970. But it was tacitly agreed that Mars would be the planet astronauts would most likely visit. In 1969, just two months after *Apollo 11,* a new report issued by a blue-ribbon panel called the Space Task Group saluted a NASA plan to keep up the momentum that had been generated by the lunar voyages. It called for a manned mission to Mars in 1981, continued lunar exploration, the construction of an orbiting lunar base, the deployment of an Earth-orbiting space station that would eventually be occupied by fifty to one hundred astronauts, and something they called a shuttle craft to ferry personnel and equipment back and forth from orbit to Earth. The new document showed the influence of von Braun, and much of it bore a suspicious resemblance to scenarios that he had outlined in his famous *Colliers* articles, written shortly after "The Mars Project."

By 1986 only the shuttle had been built, and President Reagan, like Presidents Eisenhower and Nixon before him, asked an-

The ship that finally takes humans to Mars may look like this one.

other blue-ribbon panel, the National Commission on Space, to once again map out the future. Their report called for plans to construct a space station, return to the moon, and build a lunar base and a space-transportation system between an Earth station and a Mars outpost, a kind of interplanetary trolley system that would deposit and return people and material between the two worlds. The lunar outpost would be completed by 2004; the Mars outpost by 2016. A fully operational base—really the first full-blown Martian colony—would be established by 2027.

Then, just a year later, a NASA committee, chaired by astronaut Sally Ride, completed still another report entitled "Leadership and America's Future in Space." This document said that the United States needed to take an "evolutionary" approach to space exploration. It recommended an increase in robotic exploration of Mars in the 1990s, a return to the moon, the construction of a space station and a lunar base, a piloted mission to Mars, and a human outpost by 2010.

Fascinating as they all were, the reports had not changed much during the previous thirty-five years. The Ride report, however, broke new political and bureaucratic ground. On July 20, 1989, the twentieth anniversary of the first Apollo moon landing, President George Bush appeared to take the report's recommendations to heart when he declared that the United States would set up a permanent base on the moon and mount a series of missions to Mars that would culminate in a human outpost being established there no later than 2019. At last, Mars was official. Except, now more than ever, the question became, how would we actually pull it off?

Biologists and space medical experts know all too well the monstrous technological problems of putting a human into space. A human being can survive unprotected in space for twenty seconds, at most, before he freezes and suffocates. Thus, when biologists who study these issues consider the dream of going to Mars they often find themselves collaborating with engineers—people with whom they rarely keep company—to develop technologies creative enough to make living in an utterly alien environment possible. Interplanetary travel calls for portable food, water, and air —ten pounds per day per person—ample room, protection, pleasant surroundings, and certain other amenities like companionship. Without these, humans die. In short, a journey to Mars

A transportation depot in Mars orbit. Future possibilities include mining the moons of Mars, the construction of in-space habitats, and a docking port for excursion vehicles.

Above: Soviet cosmonaut Yuri V. Romanenko exercising in space aboard MIR. In the absence of gravity, bones stop manufacturing calcium, the heart grows weak, and tendons and ligaments lose their natural resiliency. To maintain physical fitness cosmonauts exercise for two to four hours a day, six days a week.

Opposite top: Mir Quantum-Soyuz TM-3 in orbit. Soviet cosmonauts are transported to and from the Mir space station by Soyuz vehicles. Mir, Russian for peace, is equipped with a six-port docking system. It is capable of receiving the manned Soyuz crafts, unmanned cargo crafts, and modules carrying supplies and equipment.

Bottom: Mir Salyut 7-Cosmos 1686 in flight. The first cosmonauts to board the Mir space station left fifty days later to join the Mir Salyut 7-Cosmos 1686. They were the first crew to transfer from one space station to another.

requires that a little bit of Earth go along with the crew.

Soviet scientists have studied these issues more than anyone else. Mars has always held a special place in their hearts, so they have worked hard at learning how to prepare human beings to travel the vast distances necessary to get there. They first began sending cosmonauts into space for long periods in April 1971, with the launch of a mobile-home-sized space station called Salyut. The latest of these stations, called Mir, is made up of several highly sophisticated but leaden-looking modules joined together by a series of common ports and powered by bouquets of solar arrays that sprout from its assorted segments. Although not very graceful-looking, Mir works extremely well and has begun to reveal what happens when human beings are isolated in space for months at a time.

Early indications have made it pretty clear that the human organism does not adjust easily to this sort of environment, although the worst fears scientists had in the early days before Vostok and Mercury have never surfaced. There have been no lethal cases of apoplexy or lungs filling up with fluid, and, except for space sickness (the nausea, disorientation, and vomiting that afflicts half of those who fly, but which generally passes within a day or two), living and working in weightlessness turns out to be practical and actually fun, in small doses.

After a while, however, the novelty wears off, the crushing sameness of the surroundings sets in, and the body begins to experience some unusual changes. Faces puff up because in the absence of gravity body fluids float freely rather than go where they usually do on Earth. Bone, relieved of its duty to support weight, forgets its purpose and stops manufacturing calcium; tendons and ligaments lose their natural resilience; and the heart, no longer summoned by muscles to pump the blood they need to fight gravity, begins to grow weak. The immune system deteriorates, blood volume drops about 10 percent, and flavors gradually disappear as the taste buds fail. You also have to drink gallons and gallons of water just to keep from dehydrating.

All of these are tolerable . . . until the spacefarers return to Earth and the old laws of nature suddenly reapply themselves with a vengeance. After long periods in microgravity men and women who were robust and athletic before take-off return home with bones as brittle as balsa wood and hearts that race at the slightest exertion. In the early days of Salyut many cosmonauts who re-

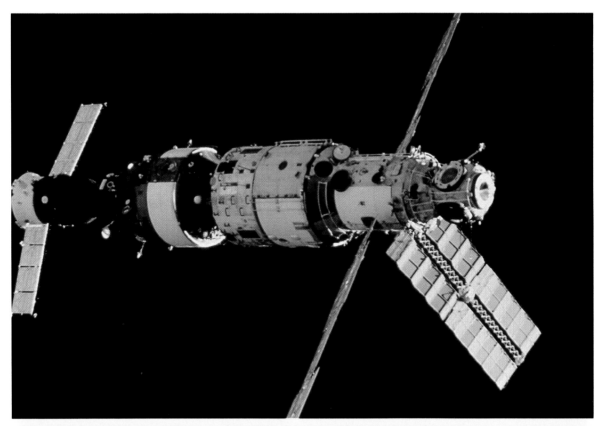

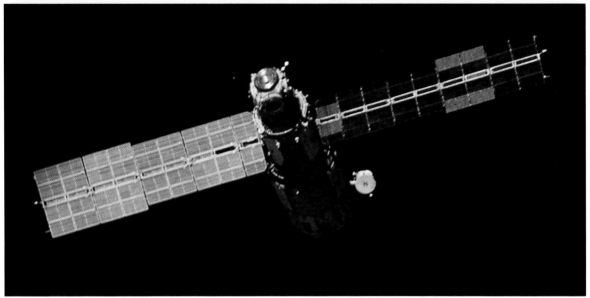

turned to Earth after extended tours in space took months to get back to normal and even then it was impossible to say how much long-term damage had been done.

Soviet scientists believe that the best way to overcome these problems is by vigorous exercise—cosmonauts are forced to ride stationary bicycles and run on treadmills for two to four hours a day, six days a week, every week they are in space.

LOVE BOAT OR SHIPWRECK: SEX IN SPACE

Cole Porter put it this way: "Birds do it, bees do it . . . Let's do it. Let's fall in love." And if human nature holds true to form, space travelers will do it too. At least that's how experts like Dr. Patricia Santy, a psychiatrist at the University of Texas in Galveston, feel. "It's foolish and perhaps dangerous to pretend that a long-duration mission will differ from any office where you have men and women working together," she recently told *The New York Times.* "Sex is a normal part of human behavior. It happens in offices. It happens in the Antarctic. It happens wherever you have males and females together."

So far the study of the need, likelihood, or advisability of sex in space remains unexplored territory. Neither NASA nor any other space agency has seriously addressed the question. But talk of assigning crew members of both sexes to the planned international space station Freedom, and renewed discussions of long missions to Mars, have made it difficult to ignore the issue any longer.

Approaches to the problem have ranged from assigning crews of all men or all women (a plan that seems to dismiss the possibility of homosexuality in space) to sending only married couples to avoiding the question altogether. Former Apollo astronaut Michael Collins recommends sending married couples to Mars. "An element of old shoe comfort," he writes in his book, *Mission to Mars,* "would be introduced by having one's husband or wife to fall back on" rather than what he calls a disastrous mixture of unattached competitors cooped up in a ship with nowhere to go.

In fact the possibility of a lovers' spat or an all-out fistfight with a lover or *over* a lover somewhere between Earth and

Mars during a multibillion-dollar mission is a chilling prospect. On the other hand, can human beings really be expected to go without sex for nine months to three years in such close quarters and under such difficult circumstances?

There are other problems associated with the issue. Contraception, for example. In a weightless environment, the more acceptable forms of contraception may not be as effective and effectiveness will be paramount. "Space may not be the best place to get pregnant," says Dr. Lynn Wiley, a reproductive biologist at the University of California at Davis. Not only could it endanger the mission, but the effects of radiation and weightlessness could harm the fetus.

But what if crew members do decide to have sex during the mission? Aside from the psychological and sociological issues, there are the problems of privacy ("space" station and "spaceship" are misnomers; space will, in fact, be at a premium). Then there is the question of actually coupling in weightlessness. Newton's third law of motion applies in a boudoir in space as well as on Earth, except with more interesting results. Lovemaking in zero gravity is likely to bring a great many *re*actions, with couples catapulting off the walls and floors, and careening into the airlocks of their tiny cubicle during the heat of passion.

The accoutrements of old-fashioned romance will also be difficult to come by. Flowers in space will be rare and champagne is out of the question unless the couple is willing to drink it from a straw and put up with the monumental case of zero-gravity-induced gas that would follow. Candlelight or cigarettes afterward? Not in space, not with all of those oxygen tanks around. In general, it's a situation that could give new meaning to Shakespeare's old phrase, "star-crossed lovers."

In addition to the workouts, Soviet scientists have come up with the "penguin suit," designed, in the words of the medical experts, to place "axial compression on the musculoskeletal system." *Suit* is a broad term; in fact it looks like a Rube Goldberg contraption woven out of elastic cords that makes standing as difficult in space as it is on Earth, a little piece of torturous engineering designed to keep the debilitating effects of weightlessness at bay. It is required cosmonaut apparel eight hours each day.

The penguin suits were invented because in the early days of

Salyut and Skylab the first men back after long tours of duty would black out when they tried to stand up, a condition caused by malfunctioning baroreceptors, sensitive pressure gauges in the body's main arteries that tell the heart how much blood to pump to the brain. After weeks in zero gravity, the baroreceptors had forgotten their job, and once the cosmonauts had returned to Earth, blood pooled in veins and arteries in the legs and abdomen, rather than returning to the top of the body where the business of consciousness is conducted. Again this wasn't fatal, but it was disconcerting. How severe would the problem become during a long mission to Mars?

Another device the Russians have invented to improve a cosmonaut's recovery time is called the Chibis vacuum suit. The Chibis essentially mimics gravity by sucking body fluids down into the legs and stomach. When they are wearing it, cosmonauts look as if they've sprouted the lower limbs of Robbie the Robot from *Forbidden Planet,* monstrous rubberized legs that reduce walking to waddling. But by wearing this suit for two to four weeks before returning to Earth, cosmonauts can be back on their feet within three or four days rather than three or four weeks.

But even these precautions do not resolve all of the nagging issues created in microgravity. American and Soviet scientists have discovered that taking diphosphates helps strengthen bones, and that vitamin D, anabolic hormones, and gallons of water loaded with electrolytes help replenish the body in other ways. To regenerate muscles, the Soviets are also experimenting with passive muscle stimulation, electrodes that zap the biceps or quadriceps or gluteus maximus with a mild electrical shock to simulate exertion. Some American scientists see these as fine interim solutions, but in the long run many pin their hopes on artificial gravity.

In von Braun's *Colliers* articles of the early 1950s he envisioned solving the problem of weightlessness with a tremendous, white-wheeled space station that created its own gravity by spinning once every twenty-two seconds. It wasn't a ship that could travel to Mars, but it illustrated how gravity could be created in space. Since then scores of futuristic designs have suggested that a spacecraft could transport a crew to Mars and back while creating Earthlike gravity along the way. Capsules connected by Kevlar tethers that rotate around one another like a pinwheel in a breeze, and three separate ships connected like children in a game of ring-around-the-rosy are two possibilities. These designs are as

The Chibis Vacuum Suit worn by Soviet cosmonauts in space essentially mimics gravity by sucking body fluids down into the legs and stomach.

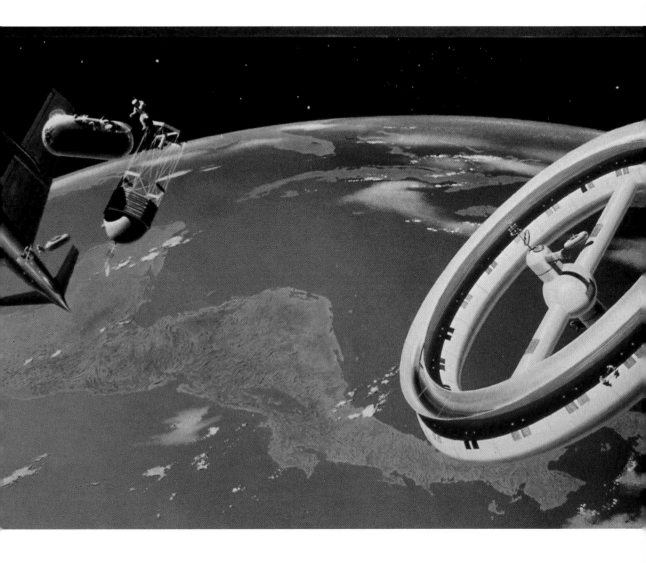

Wernher von Braun's concept of an Earth-orbiting space station. The wheel-shaped design was to supply its crew of more than one hundred astronauts with artificial gravity and all of the comforts of home. Popularized in film, television, and the artwork of Chesley Bonestell in the 1950s, it remains among the most memorable images of space pop culture. Here a shuttle from Earth approaches as astronauts assemble an Earth-observing satellite.

creative as engineering gets and, on paper, none defy the known laws of physics. But the cost of actually building dream boats of this type would be enormous. Artificial-gravity ships require completely new designs, and even if such drawing-board marvels could be translated into real ships, other scientists think they might only create more obstacles than they overcome by aggravating inner-ear problems or space sickness. The sheer complexity of them also raises serious questions. What if the system broke down 20 million miles along the way to Mars? Blowouts in space are unacceptable. The issue requires much more research, just as scores of other issues involved in a journey to Mars do.

O. Henry, the great American writer, wrote that if you want to encourage murder, lock two men up for a couple of months in a small room. This formula for butchery must have passed through a lot of cosmonauts' minds during the months they spent with crewmates in the Salyut and Mir space stations. Salyut, for example, was only six by eight by thirty feet, a very small box for such a long trip, and though no murders took place, these missions must have been terrible ordeals for the humans who participated in them.

Inside Salyut and Mir, dozens of cosmonauts have survived where there is not only no physical escape but hardly a nook or cranny for psychological relief. After a while, even with the tremendous view, the fascinating work, and the knowledge that you are among a tiny elite who makes a living laboring high above the Earth, life begins to take on all the variety of a dial tone. Privacy is nonexistent, quarters are cramped and unchanging, and the work, the experiments, the constant monitoring from Earth, and the unchanging parameters of daily life gradually become exhausting. There is no one to talk to except your crewmates and the pushy technicians in mission control barking their orders, people who have no idea what you're going through. Soon even a crewmate's most innocent behavior begins to take on sinister overtones, and simple aggravation escalates to murderous intent.

The Soviets have paid a lot of attention to how a crew gets along because that relationship is central to the success of a mission. Working together in space takes precision, concentration, and teamwork and if two competitive personalities come together under these circumstances the results could be deadly.

For example, as part of its space-station training program

Soviet scientists rigged a test that put two cosmonauts, "Vladimir" and "Nikolai," in a room at a console of instruments. Both men were told to work the instruments in order to keep them at a particular, all-important level. The men had been the best of comrades and coworkers for eight years, but to make things interesting the instruments were doctored so that each man's effort to carry out his orders was almost guaranteed to thwart his partner's. Only the closest teamwork could help them find a solution to this predicament.

For a while the two just frustrated the hell out of one another. One would operate his instruments and set his readings to zero and the other one's indicators would jump. Both men grew nervous and short-tempered. Nikolai retreated into a shell of resentment; Vladimir simply entrenched. The point of the test was to see if the two subjects could ultimately join forces to fight a common problem or if they would simply go on competing until one of them committed murder or had a stroke. Getting the readings on both dials to agree required cooperation, empathy, and human understanding; compete and the cause is lost. For Vladimir and Nikolai, the cause was lost the moment they sat down, and they had to be hauled out of the room before they strangled one another.

The Soviet scientists labeled Nikolai and Vladimir "competitive," a nice piece of Russian understatement. This competitive category was one of three groups that Soviet experts found that teams tended to fall into—the other two groups they called "congruent" and "complementary." From the interactions of each certain theories emerged.

The theory on complementaries ran against the conventional wisdom that people who are alike, who share similar values and approaches to problems, will work best together. Complementaries actually had very little in common, yet researchers found that rather than their differences being a source of conflict, these differences only made them more compatible. They didn't bicker about everything. On the contrary, they tended to carve up duties rather naturally according to their own likes and dislikes, and they balanced one another under difficult circumstances. One might be a real Tom Sawyer rib tickler who took everything in stride while the other was a taciturn nitpicker for detail. Rather than combat one another on the perfect way to attack a problem, they simply attacked different problems; or if the nitpicker started becoming a little too tense, the rib tickler would lighten up the situation with a good one-liner and put events back into proper perspective. It was stunning how well they worked together. Verbal tests with these teams sometimes showed they were in such harmony that their hearts would beat in sync!

The "congruents" were another story. The main problem with them was that they were often too much alike. An example of this sort of team was Valentin Lebedev and Anatoli Berezovoy, the two cosmonauts who spent 211 days together aboard Salyut in 1982. Both were good men, thorough and self-disciplined, but both were also industrial-strength sticklers for detail, and their similarities led to more than a few tense exchanges during their long mission. No one snapped the oxygen valve on his partner's space suit, but there were periods when the chill was so deep that mission control could sense it all the way down on Earth. In the end both men overcame their similarities because the very talents that made them similar in the first place—discipline and thoroughness—also enabled them to analyze the situation, bite the bullet, and simply will the problems away.

The United States has no versions of Nikolai or Vladimir and no Lebedevs and Berezovoys because, with the exception of Skylab, which flew three relatively short missions in the 1970s, it has never had a space station. Instead the American scientists have found a different, but very effective laboratory for studying these psychological problems: the South Pole.

At McMurdo Sound, where the U.S. Antarctic base sits on a monstrous plateau of wind-scoured ice, some members of the staff are occasionally stricken with what they call the "curse of long-eye"—the twelve-foot stare in a ten-foot room. It is the consequence, psychologists say, of spending six months in unrelenting darkness on the most godforsaken quadrant of the planet.

McMurdo and other international bases on the south polar sheet have the feel of the old gold rush outposts that once littered the Yukon at the turn of the century, a collection of simple, practical buildings hammered down to fight the elements that commonly wrack the forbidding landscape. Without continuous maintenance the howling blizzards and glacial temperatures would quickly obliterate these piddling efforts. It is a place as alien and isolated as anything on Earth.

When the big cargo planes drop off those who have chosen

to spend the winter here—mostly scientists and support personnel—the new arrivals quickly discover that this hostile environment is utterly unlike anything they have ever experienced. In the dead of winter, temperatures will sink to −100°F, the sky will turn sunless and stone black for the next four months, and they will see nothing green or unfrozen or alive anywhere. They are, as the saying goes, "on the ice," and when the last plane departs at the start of the winter season, that's it, for the duration. The Antarctic winter allows no more flights, not even for a severe medical emergency.

At first it isn't so bad. There are some beautiful sights in Antarctica. Often great auroral ribbons of light start will ripple across the sky, and when these celestial fireworks aren't in evidence the sky is translucent and un-Earthly, utterly unbesmirched by pollutants and truly stunning in a way that is unlike skies anywhere else on the planet.

Eventually, however, under the unrelenting night and endless isolation, cracks develop in the psyche. People begin to feel like prisoners. Some grow murderous or suicidal, and a few have been known to take a stroll outside...never to return. In an Australian camp, a meat-cleaver-wielding chef and a diesel mechanic armed with a fire ax once went at it like a couple of cutthroat Vikings. No one was killed but reports are that it certainly broke the monotony. At a Soviet base a Russian who had lost a game of chess to one of his comrades hacked the winner to death with an ax. (Shortly after that the word went out from mission control in Moscow to cosmonauts: No more chess games up in the space station.)

These may be unusual events, but the isolation and stress can have the most deleterious effects on an Antarctic researcher's performance. People begin to collect data incorrectly. They take longer to do their work and sometimes they don't hear or see what they should, their stomachs and joints ache, concentration wanes, they can't sleep, they grow increasingly irritable, aggressive, withdrawn, and anxious.

Psychologists call this mental retreat "general adaptation syndrome," something that happens even to the most well-adjusted people, and they find that Antarctic bases such as McMurdo make marvelous places to study what may well happen to astronauts on a journey to Mars. That trip will require coping with long periods of isolation and sensory deprivation, and will call on the crew to

Carrying its earthly cargo, a transfer vehicle separates from its mother ship and prepares to descend to Mars. Explorers in the past survived long journeys at sea that lasted for years on Earth. Will astronauts be able to survive the rigors of a voyage to Mars?

work with one another under the most claustrophobic and aggravating circumstances very far from home and family.

What scientists find fascinating in Antarctica are the novel methods people develop to meet their emotional needs. Those "on the ice" have been known to shave their heads or sport crew cuts or grow long hair kept with bows and bands. Quaint little ceremonies bubble up just to keep the mental and emotional flames alight: Some groups have mated huskies and held mock weddings. Others mount elaborate productions, complete with men made up as women.

One creative group who had endured countless screenings of Westerns, Disney movies, and pornographic films decided to cannibalize them all and spliced together a macabre little production that came out looking something like *Bambi* meets *Deep Throat*. Then they began speaking their own language based on the film. By the time the relief crews arrived in the spring, they could hardly understand what these people were saying.

If a winter in the Antarctic causes this kind of behavior, scientists wonder what might happen on the monstrously long trip to Mars? The sheer length of the mission would make Lewis and Clark wince: By some estimates it could last up to one thousand days.[6] Columbus's first round trip to the Americas was completed in less than a year. Magellan's crews, the first to circumnavigate the globe, made their voyage in thirty-six months. Captain James Cook's first scientific expedition for the Royal Geographic Society in 1768 took three years, about as long as the longest Mars missions now being contemplated. Deadly as those early journeys were, the sailors were not required to actually leave the planet or live inside a small pressurized can in zero gravity, and they could at least step off the ship onto dry land every few months. Even in the Antarctic no one lives in the stygian polar night longer than six months at a time. But for a Mars crew, even the quickest round-trip journey, made by nuclear thermal rocket and requiring a relatively short thirty-day visit, would last more than a year.

The astronauts and scientists who go to Mars will clearly have to be intrepid, well trained, smart, and above all flexible. Psychologists have found that the people who perform best under duress are easygoing men and women deeply dedicated to their work,

[6]   The thousand-day scenario would use chemical propulsion and assumes that the crew that arrives does more than simply plant a flag and return to Earth. Typically, half of this time would be spent on Mars, the other half would be divided between the outbound flight and the return home.

Previous page: A Mars Transfer Vehicle designed to take astronauts to Mars orbits the planet. This particular ship would create artificial gravity by slowly spinning its habitation modules like a pinwheel as it speeds toward Mars.

Above: Mars Direct, a mission to Mars proposed by Martin Marietta Astronautics Group. This system would use a shuttle-derived Ares launch vehicle shown here to carry payloads from Earth to Mars.

Opposite: Some scientists and engineers believe that making the long journey to Mars in weightlessness would be too dangerous for the crew. A spinning ship similar to this one could provide artificial gravity and reduce the damaging effects of living for so long in micro-gravity.

motivated, self-confident, and able to survive without the company of other humans. Their studies show that the ideal astronaut would be someone who practices the Golden Mean in all aspects of his or her life—not too intense, not too dominant, not too emotional, and not too introspective.

But in the larger picture the endeavor also requires true leaders, people with enough ambition, ego, and passion to have risen to the top of their fields and survived the difficult astronaut selection process. A person who has accomplished these feats, *and* is easygoing, is a very difficult person to find. On top of this, the Mars mission would require specialists with very specific talents and skills in geology, biology, engineering, and management. It is more than likely many skills will have to be combined in a single person like a biologist/surgeon, or a pilot/engineer. Finding astronauts who are both highly skilled and highly compatible will be tough. The best geologist for the job, for example, might not be such an easy person to get along with, but still might be the one who among all candidates could be counted on to come back with the goods.

Moreover, when eight people are put together (the generally accepted minimum number of crew for a safe round trip to Mars is eight), the dynamics of the group take on a life of its own, and the eight become more than the sum of their parts; they become a minisociety, which vastly multiplies problems. One way to address the issue is to let the group that will make the journey evolve together as a unit before they leave. This could be done by training the team together for long periods of time so that each member can find his or her emotional/social niche in the group before ever climbing aboard the ship. This approach has worked well in the military's method of building a crack unit out of the raw material of many individuals. Current plans for space station crews call for them to be broken into teams of four people who would train together in everything from flight simulators to buoyancy tanks for four years prior to serving even a single minute of their celestial tour of duty. Some psychologists like the idea of taking a Mars crew to Antarctica and simulating, in extreme and mind-rattling fidelity, what they would face during the mission itself. Whatever method is ultimately chosen, the problems will be difficult to solve and no resolution is likely to be foolproof. And this doesn't even begin to address the most basic issues of all: food, water, and air.

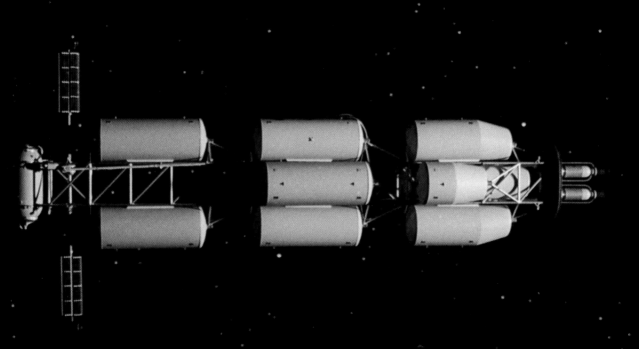

Think of a Mars-bound ship barreling through space at 50,000 miles per hour not so much as a hurtling rocket, but as a nuclear-powered organism. So many technologies, both living and engineered, will have to be joined in such seamless harmony to make this mission work that it would actually be more accurate to describe the rocket as a kind of cyborg—a six-hundred-ton, elegant conjoining of cybernetics, engineering, and living creatures, machines and humans, tool and toolmaker. It will have to be as effective and efficient as a seed with mechanical parts, a tiny, in cosmic terms, self-enclosed environment that protects and husbands the life within. It will, in effect, be alive—breathing, digesting, thinking, and communicating by a kind of radio telepathy with the planet that sent it on its way.

This very issue, the question of a self-enclosed living system, is crucial to the success of a Mars mission because scientists and engineers hope to make this traveling organism self-nourishing. Back in the sixties and seventies when Mars missions proliferated on drawing boards, there wasn't a lot of concern about how to supply a Mars crew. The assumption was they would be provisioned the same way the Apollo missions were: the ship would simply be loaded with all that the voyagers would need, the way a backpack is loaded for a long backwoods hike. No one thought much of the costs of hauling these supplies out of Earth orbit—technology would resolve those problems. But on closer inspection this approach turned out to be flawed. To completely provision even a short nine-month mission, the eight crew members would need several tons of food, water, and air—a backpack that could cost $65 million to launch at present costs.[7]

To try to resolve that problem researchers began to investigate the possibility of cultivating living crops on board a Mars ship, and they eventually came up with an invention popularly known as a "salad machine." The idea was that it would be far more practical to loft a few hundred pounds of seeds into orbit than tons of packaged and processed meat and potatoes. Moreover, it would be fresher than the freeze-dried and reconstituted food astronauts eat today.

By the 1980s evidence from the Soviet space stations was

The Controlled-Environment Life Support System (CELSS) is a NASA-sponsored project developing maximum yield hydroponic crops for use on lunar and Martian colonies. Light, temperature, humidity, airflow, and carbon dioxide levels are regulated to provide an optimal environment for the plants' growth.

---

[7]   If the ship were provisioned with every pound of water and oxygen that a crew would need for a nine-month round trip, it would amount to eleven tons, but scientists believe most water and oxygen could be recycled without a great deal of new technology. If only food had to be supplied for the entire round trip, the supplies would weigh no more than two tons.

already beginning to show that food grown in space had a salutory effect on the crew, not only because they could eat something with real flavor and color, but because they also enjoyed tending the garden. As small as these extraterrestrial patches of green were, they had become places of natural beauty among the mechanized interior of the station and the emptiness of space, and reestablished the astronauts' connection with things living, with Mother Earth herself.

At first glance this garden approach looked almost too good to be true: a natural regenerative system that performed an alchemy of sorts, converting the energy, air, and waste around it into fresh food. Provide light and minerals, small amounts of potassium, nitrogen, carbon dioxide, iron, and calcium, and seemingly out of thin air food appears. Seen this way, plants become an impressive technology in themselves.

Earth is the only known closed complex ecological system in the universe that works.[8] Mimicking it is extremely difficult. The researchers in NASA enclaves have dubbed the hardware they are testing the Controlled Ecological Life Support System (CELSS), a sealed cylinder twenty-three feet high and ten feet in diameter in which rows of wheat, fed by water and other nutrients, sprout from pipes beneath artificial light. These scientists see the myriad workings of living things as an elegant machinery, and to them CELSS is not only an apt description of their project, but of the whole planet. The question now is can it be copied, even partially, to make the exploration of *other* worlds possible?

Planet Earth has had the advantage of never having had to develop its ecology on a schedule; it has taken four to five billion years to create itself, and it has done so organically, evolving ecologies from the bottom up that grew interdependent over time. CELSS, on the other hand, is not evolution—it is engineering and it calls for a deep understanding of what it takes to quickly invent a system that can regenerate itself.

Progress has already been made on this front. At the Kennedy Space Center in Florida, for instance, scientists have successfully grown compact rows of crops in mineral-rich solutions of water, and by varying the atmosphere, mineral mixture, and light intensity they can grow wheat more efficiently than Dakota farmers, and

Konstantin Tsiolkovsky's drawing of a dwelling or "hothouse" in space. He foresaw the advantages of enclosed living systems that mimic Earth's biosphere.

---

8   There are some very small enclosed living systems that apparently work. For example, at the University of Hawaii, biologist Clair Folsome has kept colonies of bacteria growing in sealed laboatory flasks since 1967.

cultivate potatoes with far better yields than anything ever planted in Idaho or Maine. Recent estimates indicate that about twelve square meters of wheat would feed one astronaut on a long journey. A little work in the ship's galley and various crops could be transformed into pasta (wheat) with homemade marinara (tomatoes, herbs, and spice), a fresh salad, and hot sliced Italian bread.

Soviet scientists have been working in this field longer than anyone. A Russian mineralogist named V. I. Vernadsky defined the concept of living systems early in this century when he described Earth as a "biosphere." In 1912 Tsiolkovsky himself wrote, "In order to ensure a food supply for a man during a flight in a spaceship, he must take with him various plants that will purify the air and bear fruit." After years of frustrating efforts in their Salyut space stations, the Soviets eventually succeeded in raising wheat, oats, peas, carrots, dill, and a few other crops in a contrivance called Phyton.

But these experiments did not really tackle the most complex problem, the issue, as scientists put it, of "closing the loop." A closed environmental system means that everything alive inside of a Mars-bound ship, both plant and animal, must absolutely depend on every other organism on board for its survival, exactly as on Earth. This means that all waste, leftovers, organic garbage—whatever refuse that is created, both wet and solid—must be completely recycled through the system. Plants will serve as the ship's lungs—a healthy crop of wheat can supply more than enough clean oxygen for a crew of eight—and can also be enlisted to purify and recycle the ship's water supply. Among the many laudatory side effects of photosynthesis is that it distills water by releasing it in the form of vapor, which can then be condensed and used for showers, tea or coffee, or drinking water. This would eliminate the need for any fancy on-board water treatment system by simply replicating the same processes that do the job naturally every day on Earth.

But despite some early successes, scientists don't know how well crops will grow in zero or low gravity for long periods of time; their yields could deteriorate over several generations, or their inner workings might become scrambled as they do in humans. Solar and cosmic radiation may also be important factors, and if radiation damages humans, it may wreak genetic havoc with seeds as well.

In the near term tests in space stations, or on moon bases,

A full-scale mock-up of a portion of an aerobrake in an underwater test tank. The aerobrake, which may be used in future space travel, is being developed by NASA.

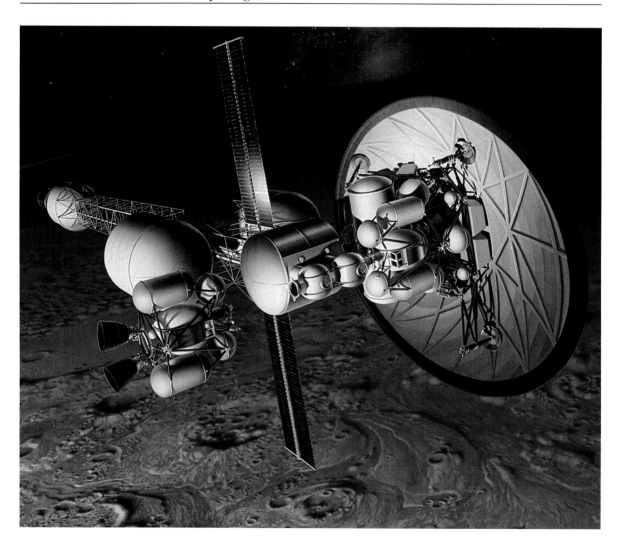

Artist's rendering of a Mars-bound
ship with an aerobrake.

will have to determine the real chances of successfully growing crops in space. If they succeed, they will have gone a long way toward making it possible for the first humans to safely reach Mars. Of course, getting there is one problem, exploring it is another, and its exploration raises some of the toughest technological, scientific, and philosophical questions we have ever faced.

When the first Mars-bound crew finally reaches its destination, the ship will swing into orbit to dock with a cargo ship that arrived a year earlier, loaded with equipment and supplies for the trip to the planet's surface. Following this little celestial ballet, the Mars lander will be released with three humans inside. The lander's aerobrake, a curved palette the size of a small skating rink, will be deployed, its broad surface slowing the ship's entry through the thin Martian atmosphere. The brake will be jettisoned and retro-rockets will softly deposit the vessel on the planet's surface.

The moment the three humans inside the vessel step out onto the rusted surface of Mars the true desolation of the place will hit them. No place on Earth, with the possible exception of Antarctica's dry valleys, is as stark. Even the most barren stretches of the Mojave desert have an arthritic Joshua tree or two clinging to the rock, but here it will be utterly dead—not a bird, lizard, or cricket, not even the green haze of vegetation on mountains in the distance.

But Mars probably won't feel dead in the same way that the moon felt to the Apollo astronauts; even before going there no one ever expected the moon to be anything but lifeless. But on Mars there will be a faint echo of the living, an almost imperceptible sensation that life once existed there. How much of that feeling is real and how much is the piled up, psychic baggage packed into our subconscious by the fantasies of Lowell, Burroughs, Bradbury, and Wells will be difficult to know. But there will be other echoes too. The Martian sky, for example, isn't blue like Earth's, but pink because of the ruddy refracted Martian dust that fills it, and the sun on Mars will be more feeble, only half as bright, and two-thirds as large as it appears from Earth—a twilight sun—but it will be there rising in the east and setting in the west, and there will be something comforting about the pace at which it marches across the sky—the Martian day is only forty minutes longer than Earth's. The experience will undoubtedly seem surrealistic, a mix of the familiar and the unreal.

Following their long journey from Earth, the Mars exploration team's ship would dock in orbit over the red planet with a cargo ship sent ahead from Earth a year earlier. Could Mars, slowly spinning below, reveal secrets about the origins of life that were erased long ago on Earth?

Despite their state of mind and the stresses of their long, weightless journey, the astronauts will have precious little time to acclimate to Mars's gravity, which, although only one-third Earth's, will suddenly make them feel leaden. They will unload and erect their Spartan living quarters, the mission's scientific and communications equipment, and a rover from the belly of the lander. The rover will be slow, even ponderous, and the explorers will not find themselves zipping over Martian hill and dale like the Apollo astronauts did in their little balloon-tired lunar buggy. Nevertheless, there will be a lot of ground to cover; the explorable landmass of Mars is just about the same as Earth's because, even though Mars is a smaller world, most of Earth's land is under water. It will be, as one scientist has put it, a little like landing in the Mojave desert in a hi-tech Winnebago to explore seven continents at five miles per hour.

But speed might not be all that critical. The first Mars mission will probably not attempt to travel great distances. Between the Mars Observer spacecraft, due to arrive in 1993, and work done by proposed robotic rover missions, astronauts will undoubtedly have a precise itinerary before them. By the time their mission is launched, all branches of science will have prepared a renewed barrage of inquiries about Mars, and it will be the explorers' job to begin the planet's cross-examination. Geologists will want to know more about Mars's huge volcanoes, and demand a firsthand look at the Valles Marineris. How does a planet make a scar in itself that huge? Climatologists will hope to discover the secrets locked in the layered Martian polar caps, which might finally unveil the mystifying behavior of the sun and its effects on Earth over the past four billion years. Above all, science will want to answer the big question: Is, or was, there life on Mars?

On the Martian equator lies a place called Hebes Chasma, a box canyon snuggled in a northern corner of the Valles Marineris. This ancient lake bed is the site Chris McKay, the scientist so fascinated with life-forms in Antarctica, would target as the first place to send astronauts if he could have his way. If Mars remained warm enough in its early history to have created large bodies of water, then Hebes Chasma would have made an excellent cradle for early life. Volcanic vents or mineral-rich springs may even have existed nearby, like those that surround Old Faithful in Yellowstone National Park. Even after the planet cooled and its mean temperature dropped below freezing, McKay and others have surmised that life

This Mars rover called a "walking beam" is being developed for future unmanned missions to Mars. It could collect soil and rock samples over several months.

may have found a way to survive there anyway. On Earth the mean temperature of Antarctica's dry valleys is −4°F. Once, not long ago, scientists believed that life could not possibly exist in such a place, but it does, and the same situation may be true on Mars as well.

Either way, McKay believes that the lake bed in Hebes Chasma would make an ideal place to search for the first evidence of Martians. Explorers might not hope to find anything as large as a dinosaur bone or the remains of an alien plant, but they may find layered mats, evidence of ancient microbial communities that died off billions of years ago but remained perfectly preserved in a primordial ooze now frozen hard as marble. Assuming a rover were rigged with a drill, it would be a simple matter to sink it into the Marsscape, draw out a core sample, and see whether life had ever existed there.

As far back as 1962, Carl Sagan and Nobel Prize–winner Joshua Lederberg suggested that future explorers might find fossils of Martian life imprisoned in the planet's permafrost. In the midst of the many Martian-life theories that followed, the permafrost idea never really garnered much additional attention, especially after Viking. Then a couple of years ago, Imre Friedman, the same man who found the strange yeast in Antarctic rocks, met a group of Soviet biologists from the Institute of Soil Science and Photosynthesis who had been doing research for the past twenty years in the permafrost in Eastern Siberia. There, in tundras near the very same gulag where Soviet rocket master Sergei Korolev had been held prisoner during the 1930s, these biologists had found something astounding in the frozen ground: countless microbes that had been frozen solid for three million years! Furthermore, they found that if you scraped a few of these organisms off the ice and placed them in a petri dish with a splash of water, they instantly came back to life, as if they had simply been awakened from a long nap.

It had never occurred to the Russian scientists that their discovery could have anything whatsoever to do with extraterrestrial life, but Friedman and Chris McKay immediately saw tremendous possibilities. If these microbes could survive frozen at 14°F for three million years, and then simply be revived by adding water, could Martian microbes have survived at −94°F for three *billion* years? And could they be brought back to life in the same way? Possibly. Microbes on Earth have been frozen down to −148°F

The Mars Observer. Scheduled to be launched in 1992, the Mars Observer will orbit and investigate Mars in more detail than any probe since the Vikings in 1976. It was designed to be a small, low-cost reconnaissance vehicle.

Following pages: Two astronauts explore Noctis Labyrinthus in the Valles Marineris on a misty early morning some time in the future.

U. S. GEOLOGICAL SURVEY
FLAGSTAFF, ARIZONA

and then revived. In fact, Friedman and McKay have wondered whether the organisms in Siberia are truly frozen or simply living at a colossally sluggish rate. Under normal temperatures these microbes might live ten years, but down in the tundra, their movements would be so glacial, so immeasurably torpid, that their life spans might last billions of years. They could even, ever so slowly, be multiplying in the ice.

This could mean that life-forms might actually still exist on Mars, and if they do, McKay and Friedman wonder, could they be revived? And if so, could the entire planet eventually be brought back to life?

Even if no microbes managed to survive, they might still remain perfectly preserved in the Martian permafrost, a possibility that raises fascinating questions. What would a Martian microbe look like? Would it bear a resemblance to its Earthly cousins? What of its genetic structure? Would it even have genes? And if not, how would it reproduce?

If such a discovery were ever made, it could help unlock mysteries about the origin and nature of life. It would instantly revolutionize evolutionary biology, astronomy, and all planetary sciences, upend whole philosophies, and most of all shatter our notions of where we stand in the scheme of things—the very question that leads us to Mars in the first place.

Furthermore, such creatures might unlock the very secret of life in a way we've been unable to resolve on our own planet. The story of how life originated on Earth has been erased. The oldest rock on the planet is 3.8 billion years old, and nothing of Earth's original surface remains; it has all been buried, melted, or renewed, as has any evidence of how the first cells formed. We only have the remnants of microbial mats found in Australian rocks that are 3.5 billion years old—life-forms that were already very sophisticated by the time the rocks came into existence.

Unlike Earth, Mars has not completely resurfaced itself; it cooled more rapidly, and large portions of it remain much as they were more than 4 billion years ago. The river channels that border the Valles Marineris are older than any single geological formation on Earth, and could very possibly hold some clue to how life could arise out of inanimate rock, water, and chemicals.

A Mars composite photo based on images returned by several unmanned space probes. The rocket, complex guidance systems, and new imaging techniques have opened the doors on new worlds and new experiences.

If we do not find life on Mars, however, we will almost certainly transplant it there. Our plans to build miniature biospheres mimicking Earth that astronauts can take with them into space suggest that we cannot survive very long outside of the company of other life. For short periods of time we could backpack to other planets, but over the long haul we will not survive journeys to other worlds unless we learn to bring a slice of Earth along with us. Nor will we successfully colonize Mars unless we also learn to establish the same kind of closed living system on that world that we enjoy at home.

Early efforts to accomplish this feat would have to be simple, but eventually they would allow the first Mars bases to grow into self-sustaining outposts and then colonies. More crops, more efficient recycling, and undoubtedly more surprises would evolve. Plants might be genetically engineered so that they could survive with less water, or more solar radiation, or in colder temperatures. Gardens, farms, and ranches could flourish beneath Teflon or Kevlar domes that protect humans, animals, and plants from the sun's radiation and Mars's cold. Solar satellites, wind turbines, and solar power arrays could supply energy. Water could be drawn from Mars's permafrost or, ironically, from the same polar caps that Percival Lowell thought were supporting the dying Martian race he first imagined nearly one hundred years ago. According to plans being put together by Obayashi, a Japanese manufacturing company, Mars could be the home of 150 humans living in a fully self-sustaining habitat that would cover more than 500,000 square yards of Marsscape by the year 2057—the centennial of the *Sputnik* launch. Other proposals have envisioned saving enormous sums of money by fashioning brick, mortar, and plastic structures six stories high out of Martian soil, an approach that would eliminate the need for shipping tons of metal and inflatable modules from Earth, millions of miles away.

From these colonies the exploration of Mars would continue and life would expand. Humans and robots would set out for the poles, the canyons of the Valles Marineris, and the lava flows of Olympus Mons. Eventually we might undertake the terraformation of Mars in an effort to render the entire planet alive and Earthlike, a project that would call for creating an atmosphere that is denser, a climate that is warmer and wetter and safer. This in turn would require pumping the Martian atmosphere full of greenhouse gases —the same gases that, ironically, are polluting our own planet right now—in order to raise its temperature and turn its ice into water. Though this process could take 100,000 years, the planet

could be made to support insects, plants, and microbes within just a few decades. It would be an immense project, one Lowell's Martian engineers would undoubtedly appreciate.

Such an effort would eventually make humans a multiplanetary race, and the harbingers of life. But perhaps on some scale too difficult for us to comprehend right now, our explorations of Mars really represent nature's way of steadily expanding to new habitats. Humans believe they are in control, but maybe our toolmaking abilities, which enable us to build complex machines like spaceships and systems that can protect us and carry us into space, are simply clever inventions that life on Earth has developed in order to carry it to other worlds. Lynn Margulis of the University of Massachusetts at Amherst, and one of the most respected, if controversial, microbiologists in the United States, has pointed out that throughout time living things have invented all manner of marvelous methods for finding new ecological niches. Propagules, for example, are clever contrivances in nature that allow organisms to propagate. All seeds are propagules. So is a kernel of corn, a spore, and an acorn; and, it could be argued, so is a spaceship bound for Mars, with its regenerative systems and human beings inside.

But whether we are the agents of our own destiny, and whether or not Earth will someday reanimate Mars and generate a copy of itself, our exploration of the red planet will represent a sort of homecoming. Long ago Mars held all of the constituents necessary for genesis; the evidence is still written all over the planet's surface. It was warmer, wetter, and more active, but when it cooled, those constituents were banked in its crust. It locked water up into ice, and trapped oxygen and life-giving nutrients into the rock and, having lost its thick atmospheric blanket, it released its ancient warmth to space. With a little coaxing, however, the processes that placed these constituents in planetary escrow might be reversed.

Maybe there is a certain sense in this. After all, if life had once established itself on the planet, wouldn't it seem reasonable that life, even life dispatched from another world, would cause it to flourish there again?

A view of Mars after it has been terraformed, a process that would re-create the planet in the image of Earth. Transforming Mars on this scale could take a hundred thousand years, although certain early changes could allow microbes, plants, and insects to live in the open only decades after the process is begun. Making Mars habitable means increasing Mars's temperature by raising its atmospheric pressure, which would in turn melt water in the ground and at the planet's polar caps. More oxygen would create an ozone layer to protect life on the surface from ultraviolet radiation. Should this ever happen, life will have succeeded, in the form of the human race, in creating a new Earth-like world in another part of the solar system.

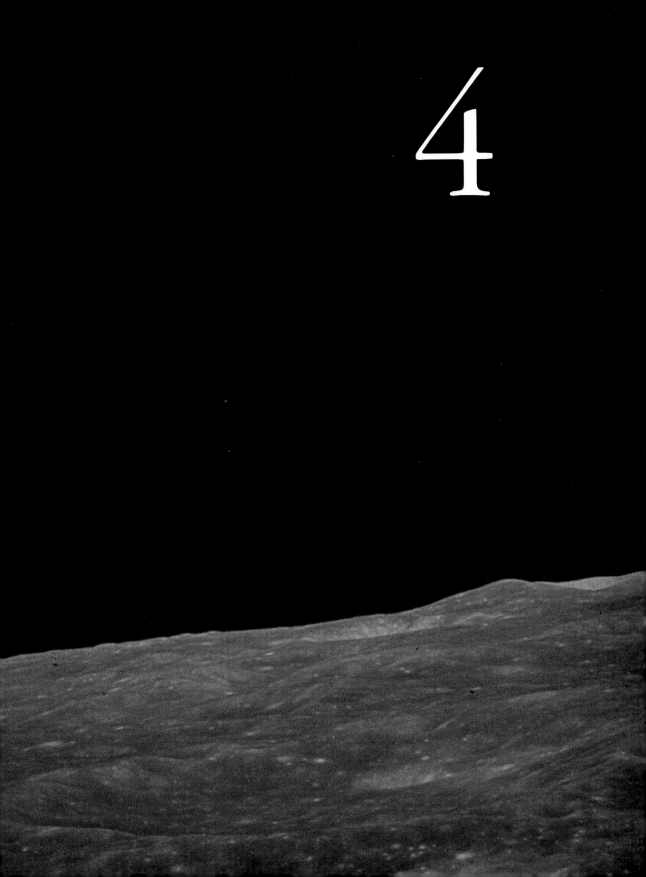

"Human history becomes more and more a race between education and catastrophe."

—H. G. Wells

Mission to Planet Earth

Above: This Babylonian clay tablet is the earliest known map of the world, rendered around 600 B.C. The Babylonians saw themselves as the center of a world bordered by an unconquerable body of water, beyond which there was nothing.

Left: Asher B. Durand's painting "Progress." Early in the Industrial Revolution, a harmony between nature and industry was the ideal.

When we went to the moon for the first time, we discovered a new world: our own. We were reaching out that day, leaving the orbit of our home planet and bent on reaching an unfamiliar destination. For three days the *Apollo 8* module traveled away from Earth, and then on December 24, 1968, the capsule and the three humans inside it fell into a dark orbit behind the moon. As they circled back toward the Earth, they were suddenly stunned at the sight that rose over the barren lunar surface. There in the appalling blackness hung this tiny, gemlike world, Earth, a place that had somehow miraculously blossomed in an otherwise utterly sterile parsec of the cosmos.

Seeing our own planet rising above the moon's horizon is arguably the most profound and unexpected gift of the Space Age. On an almost subliminal level we immediately sensed that we do not bestride the planet like a colossus, but are a part of it. Yet we had to leave Earth in order to see it this way. Looking back across a quarter of a million miles of utter emptiness that night, Earth appeared fragile and miraculous, no longer the immense world we had been struggling for so long to dominate.

If you look at a succession of world maps drawn over the past twenty-five hundred years, you can recognize the human effort to overtake and comprehend planet Earth, long before we had any concept of what a planet was. It is like watching a blind man fumbling to make out the edges of an enormous room. The first known map of the world was created by a Babylonian six hundred

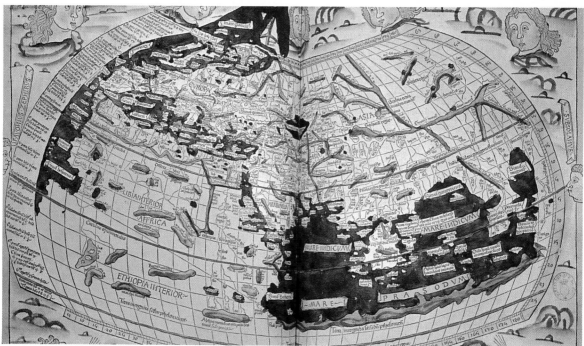

years before the birth of Christ, and it was not exactly a wonder of objective cartography. The mapmaker had sketched it on a clay tablet as a flat disk of land surrounded by an ocean and certain mythic islands. The disk was dominated by the city of Babylon, with Assyria to the east, and Chaldea to the west, and that was the extent of its detail. It says much more about the way Babylonians viewed themselves in relation to the rest of the world than it does about the geography of the world itself.

Over time, however, each successive map illustrated a broadening circle of knowledge—a shoreline added here, an elbow of land there. Continents and oceans appeared. By the second century before Christ, Eratosthenes, an Ionian scholar, reckoned that the world was round, and in the early sixteenth century, Amerigo Vespucci, an acquaintance of Christopher Columbus, declared that the shorelines of recently discovered lands were, in fact, parts of previously unknown continents—the New World. As the maps grew increasingly accurate, they signaled our success as explorers and as a species, evidence that Western civilization was spreading into all corners of the globe, growing fast and gathering knowledge. Today we are everywhere, and our success as a species has made us one of the dominant forces of nature.

When the first human artists were painting the caves of Lascaux, France, sixteen thousand years ago, there were perhaps only a few thousand modern humans on Earth. In 1750 there were a half billion of our kind, and by 1950 our numbers had grown to two and a half billion. Today, a mere forty years later, more than five billion human beings rely upon Earth to fulfill their needs, and the population still rises while the planet's resources shrink. This dilemma lends a new urgency to our old efforts to comprehend planet Earth. We must greatly improve our view of Earth, learn not only the dimensions of its seas and landmasses but also the intimate workings of its processes and our role in those processes. If we don't, we may destroy ourselves.

It is not the simple growth of our numbers that has placed unprecedented stress on planet Earth. The skillful methods we have devised to use the planet's resources have also accelerated the effect. We are dedicated toolmakers, a talent that has enabled us to both open up the world and draw upon its gifts, as well as leave it and see it anew. All life draws energy from its surroundings, but since the days of *Homo habilis* and his stone tools we have ex-

Above: The cave paintings at Lascaux were created by people known as the Magdalenians between eighteen thousand and eleven thousand years ago. The delicacy of the paintings, the use of perspective, and the imaginative incorporation of subject matter with bumps and crevices in the rock have awed even great artists of the modern world.

Opposite top: This image of the Earth is a composite image based on information from many Earth-imaging satellites. It shows both the ocean floor and Earth's land masses.

Opposite bottom: The Ptolemeic projection. Although there are no known maps by Ptolemy in existence, his projection was a significant event in cartography because it marked the first effort to illustrate a global image of the Earth on a flat surface. Despite its misrepresentations, this is one of the earliest attempts at an accurate, scientific depiction of the world.

ploited and reshaped our environment with unusual success. In late-eighteenth-century Europe we made technological leaps that suddenly enabled us to do this on a radically larger scale, a convergence of events we call the Industrial Revolution.

The signal advance of this revolution came in 1769 when a Scottish inventor named James Watt perfected the first efficient steam engine, an amazing assemblage of pumps, gears, and levers for converting heat energy into mechanical energy. Soon after creating it, Watt went into partnership with an entrepreneur named Matthew Boulton, who quickly foresaw the enormous potential of Watt's work and built a steam-powered factory in the English market town of Birmingham. In 1776 when a biographer came to visit Boulton and asked him what purpose his factory served, Boulton declared, "I sell here, sir, what all the world desires to have—power."

That, in a nutshell, defined the Industrial Revolution. By stoking enormous steam engines that could do the work of scores of men and women, Boulton suddenly redefined power and ignited the world. His Birmingham factory initiated the proliferation of machinery in a society of farmers and herdsmen, and accelerated the rate at which the ingredients of the planet could be placed into the service of the human race. Rich seams of coal were exploited, releasing the energy of the decayed trees, plants, and other organisms banked beneath the surface of Earth. For better than two hundred years we have been drawing energy from the planet in this way—not only from gas, oil, and coal but from wood, crops, water, and animals. Today we move fabricated or harvested parcels of the Earth around like ants carry the grains of an ant hill; we and our machines are drinking, breathing, and exhaling on a worldwide scale. At issue now is whether we are going to eat ourselves out of house and home, or whether we will use the new tools developed since the days of Apollo to understand the complexity of our world and search for ways to live in harmony with it.

Science has traditionally divided the study of the Earth into a series of disciplines: geology, oceanography, climatology, biology, meteorology, each artificially designed to render Earth's behavior comprehensible. But the planet strains under these harnesses. Astronauts who have looked at Earth from space understand why: From their unique viewpoint they have seen that the planet is not carved up according to disciplines or imaginary political borders; it is completely seamless.

Apollo 9 astronaut Rusty Schweikert recalled that on his earlier orbits of Earth he identified at first only with places associated with personal memories—Houston, his home, and Los Angeles and Phoenix, where he had flown as a pilot. On later orbits, as the land and sea below became more familiar, he began to identify with North Africa and other large parts of the world. Soon he found himself identifying with the entire planet; it had all become one place, and he simply couldn't imagine borders any longer. He knew, he said, that in the Middle East people were killing each other over some imaginary line, but he couldn't see that line. He remembered wishing that the people on both sides of the fighting could see what he was seeing so they could ask themselves: What is worth so much hatred?

Shuttle astronaut John-David Bartoe also remembers seeing an undivided world. "As I looked down, I saw a large river meandering slowly along for miles, passing from one country to another without stopping. I also saw huge forests, extending across several borders. And I watched the extent of one ocean touch the shores of separate continents." From this perspective it seems so obvious that the world is a place, despite its size, where nothing and no one is really separated from anything or anyone else.

Like the view of Earth from the moon, these revelations are more emotional than intellectual, but they mirror the scientific reality that Earth is one immense integrated system. Vladimir Vernadsky, the Russian mineralogist who developed the idea of the *biosphere,* was the first scientist to recognize this. He was fascinated by the way chemical forces broke and reshaped the Earth's crust. He realized that over time this chemistry dissolved whole mountains and gradually deposited them into the sea. Atoms moved up and down and around the crustal skin of the planet, forming new minerals and chemical compounds, and molecule by molecule the planet was constantly remade.

In the process of these studies Vernadsky created the science of geochemistry. But even this new science couldn't fully explain all the mysteries he saw in the Earth, because on closer inspection he realized the processes were not chemical alone. Life, from the smallest microbe to the largest forests, was creating minerals—water, carbon, oxygen—and these molecules migrated in grand cycles everywhere on the planet, like blood through a body. But

Following pages: The Straits of Gibraltar, where the Mediterranean Sea flows into the Atlantic Ocean. Africa lies to the south, Europe to the north. From space all boundaries are invisible.

Vladimir Vernadsky coined the term *biosphere* and was the first person to realize that Earth is an integrated living system.

they were able to do this only because living organisms made it possible, an insight that led him to conclude that life is deeply connected to all aspects of the planet's operation.

In 1926 Vernadsky fully articulated his grand theory in a series of lectures he delivered and later published in France under the title *La Biosphere*. Earth, he told his audiences, is not simply a rock upon which living organisms happen to be growing, it is a vibrant, integrated zone in the sterility of space where a planet's nutrients met with the forces of a star and somehow kick started the life that had become central to its operation.

In one of these lectures he anticipated the Space Age, and imagined what astronauts and cosmonauts would later witness for themselves: "The surface of the Earth, seen from the depths of infinite celestial space, seems to us unique, specific and distinct from that of all other heavenly bodies. The surface of our planet, its biosphere, separates the Earth from its cosmic surroundings."

Vernadsky's theories marked the beginning of a more integrated scientific view of Earth's behavior—but only a beginning. The scientific mainstream continued to carve the planet up among a corps of artificial disciplines and was slow to see Earth as a world seamlessly connected by the life upon it.

Then in the 1970s another maverick scientist, a quiet British atmospheric chemist named James Lovelock, stepped forward, independently of Vernadsky, with an even more outrageous theory. Life, Earth's biosphere, he said, is not only integrated into the planet's clockwork cycles, it is controlling them. He had developed this controversial idea in the 1960s while working for the Jet Propulsion Laboratory as an atmospheric chemist on the Martian Viking probes, which were designed to check Mars's atmosphere for signs of life. But before these probes were even launched, Lovelock had concluded that they wouldn't find a single living organism because Mars's atmosphere is, as Lovelock put it, chemically stable and unreactive. In other words, given the chemistry of the planet and the forces at work upon it, the constituents of its thin atmosphere are exactly what one would predict.

Earth's atmosphere, on the other hand, makes no chemical sense at all: It is dynamic in ways that physics and chemistry do not predict. Its gases are not only different from those of Venus and Mars, the nearest planets, but different from that which Earth's chemistry itself should have produced. The Earth's atmosphere should be mostly carbon dioxide, like Mars, and nitrogen and oxygen should be nearly nonexistent. Instead, carbon dioxide is a mere trace gas, the planet is swimming in nitrogen, and nearly a quarter of its atmosphere is made up of oxygen—one of the rarest, most reactive gases in the solar system. From an atmospheric chemist's point of view, this is like visiting the moon and finding a fully assembled Boeing 747 sitting in the Sea of Tranquility. It would not be likely that a jet could have spontaneously formed out of the minerals of the moon. Earth seems to defy a basic law of the universe—the second law of thermodynamics, or the law of entropy, which stipulates that in the natural course of events the constituents of the universe—everything from stars to molecules—are prone to fall apart, slipping toward equilibrium. But on Earth, events seem to be running contrary to these principles. The culprit, Lovelock maintains, is life, a force unique in the universe because it is self-organizing. Life is, in fact, a new kind of energy—a regenerative form—and if you place the variable of life in the equation, then Earth makes sense.

Few argued with Lovelock that life plays a key role in Earth's behavior. By this time even the most conservative scientists, spurred by space-probe observations of other worlds and the view of Earth itself from space, had come around to Vernadsky's theory about the biosphere. But Lovelock had gone farther than Vernadsky, holding the highly controversial view that life is more than simply important to the operation of the planet; it actually regulates Earth's cycles, adjusting the whole global machinery to protect itself and ensure that the planet remains habitable in an otherwise uninhabitable sector of the universe.

Lovelock dubbed this concept the *Gaia Hypothesis* in honor of the Greek goddess of the Earth. Most scientists are skeptical of Gaia because they believe it smacks of intention and purpose, a concept that runs counter to Darwin's theory of evolution. Microbiologist Lynn Margulis of the University of Massachusetts, however, who worked closely with Lovelock as he developed the Gaia hypothesis, claims that it is absolutely consistent with Darwin's theories. Since the first cells developed on Earth more than three and a half billion years ago, she notes, life has grown increasingly diverse and complex in the face of formidable odds. Periodically meteors and comets have collided with the planet, nearly wiping out everything alive upon it. Climate and continents have shifted, the sun has grown steadily warmer, magnetic poles have flipped, the planet has shuddered under wave after wave of earthquakes,

Lynn Margulis of the University of Massachusetts first discovered that eucaryotic cells evolved when procaryotic cells banded together in a process she calls symbiogenesis. She now believes that symbiogenesis drives evolution and unites Earth in a global living system.

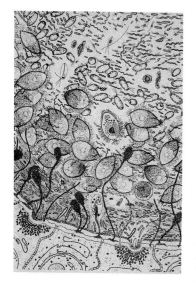

ice ages, and volcanic explosions. Yet life flourishes. As a recently forged link in this living chain, we ourselves are evidence of its power to adjust to change. How could all of this be if there weren't some very flexible force at work regulating the global system?

For Margulis the heroes of this remarkable success story are the very first forms of bacteria, creatures that first arose on Earth more than three and a half billion years ago. For the first 75 percent of the planet's history, these microbes were the only life-forms in existence, and although we tend to think of them as simple, Margulis points out that these tiny creatures did all the truly difficult evolutionary work. It was not the more complex forms of life that first developed predation, social organization, photosynthesis, motion, and sex—not to mention the greatest accomplishment of all, genesis itself; it was a tiny bacterium called a procaryote. Together these microbes transformed Earth from a molten rock orbiting the sun into the blue, shimmering world we see from space today.

A vast community of these ocean-dwelling microbes began releasing the oxygen that we breathe, and the ozone layer that protects us from the sun's radiation. Lacking these inventions, no life-form could ever have made it safely to land or gone on to evolve into trees, dinosaurs, apes, or humans. Without microscopic life, the planet would literally dry up and blow away.

All of this has led Margulis to propose another controversial theory called symbiogenesis, which holds that many different kinds of life-forms are physically connected, a belief that links their behavior with certain mysteries that the Space Age has begun to reveal. The evolution of life on Earth, she believes, has been driven by a series of biological mergers among bacteria that ultimately enhanced the survival of all participants. Alone each cell might have limped along, but by joining together in larger groups they became more powerful and efficient than any one of them could ever have been on its own. In time the borders of these microbial confederacies blurred, and out of their symbiosis emerged an altogether new living thing. Gertrude Stein wrote, "A rose is a rose is a rose." But for Margulis a rose isn't simply a rose; it is an evolved network of microbes, millions of different kinds of cells that have joined forces to do what a rose happens to do. The only thing that makes an organism unique is the way it orchestrates this basic life-form. Certain cellular alliances have resulted in plants whose blooms we call roses, certain others have

In the inside of a termite hind gut microorganisms work to digest wood.

blossomed into humans like Albert Einstein and Charles Darwin.

Margulis now believes that symbiogenesis has also created the very foundation of the manner in which all of planet Earth runs. If it is the driving force in evolution, as she believes, then Earth's cohesiveness as seen from space is more than a poetic image: It is a biological reality. The planet is a colossal collusion, intimately connected, cell by single cell, and individuality is an apparition. It creates a single organism of the entire world, with different groups of cells in the oceans, forests, air, and land working together under innumerable guises to keep the whole planetary apparatus up and running.

If Margulis is right, if we ourselves are literally part of a grand cellular congress called to order some three and a half billion years ago, then the most recent images of Earth from space raise a deep and unsettling question. Are we a constructive or a destructive force within the planetary corpus? Is the Space Age itself a new method of symbiosis, a monumental survival technique in which our toolmaking abilities are sweeping evolution in new directions? Or is Earth, as the oceanographer Roger Revelle once wondered, suffering from an epidemic of human beings?

We humans have wreaked a lot of havoc on Earth, but in 1968, when *Apollo 8* first circled the moon, we did not understand the full extent of the damage we were doing. We never imagined that between then and now we would bring the panda, elephant, rhinoceros, bald eagle, and uncounted thousands of other species to the brink of extinction. Commercial whaling was profitable and practiced worldwide. Carcinogens were freely pumped into the atmosphere and freely scattered across continents and oceans for all to breathe and ingest. At that time it was still inconceivable that we could fish the oceans lifeless, but today some sections of the world's seas are being stripped clean.

We are devouring the planet in great chunks, plundering it of the very life that brought us forth. The Apollo astronauts couldn't see the evidence of this, but today's astronauts can, and without the aid of any special instruments. Even from two hundred miles high, the Earth looks a little more polluted and opaque than it did in 1968, and it is showing the effects of rapid ecological damage.

Mike Helfert, a science investigator at NASA's Earth Observations Office in Greenbelt, Maryland, has trained dozens of astro-

Mike Helfert of NASA has taught dozens of astronauts to photograph Earth from space. The program has yielded over 140,000 photographs of Earth taken in space.

Lake Chad, north central Africa, June 1966.

Lake Chad in September of 1988. Despite heavy rains that occurred in this area immediately before the picture was taken, the lake is at its lowest level ever seen by astronauts. Since the 1960s the level of the lake has decreased by 90 percent.

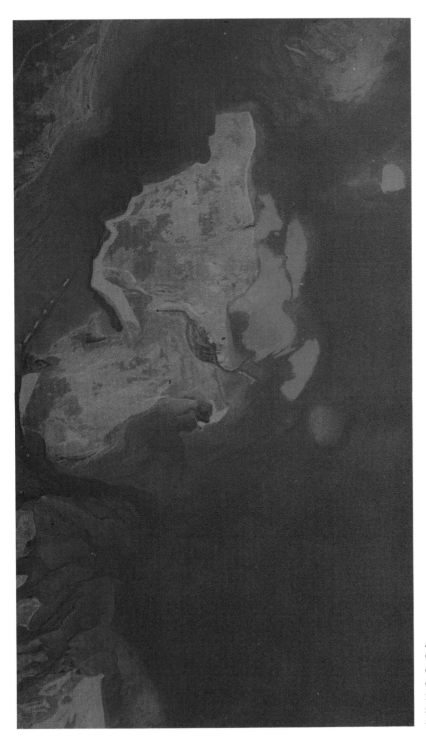

An oil spill in the southern Persian Gulf, May 1991. The source of the oil was not determined, but if it originated from Kuwaiti offshore oil fields damaged in the Gulf War, as suspected, the slick traveled at least 460 miles.

nauts to photograph Earth from space. This special form of orbital portraiture has yielded over 140,000 different images of Earth from space. Though these pictures don't provide the precision or range of digital satellite information, to Helfert they are unparalled for their emotional power. He considers astronauts wonderfully intelligent sensors who rarely miss photographing an amazing image when they see it. Many of the photographs they have returned—dust storms whipping across the Pacific or tornadoes roaring on their sides like fallen tops—have redefined our view of the way Earth's climate works.

Today, each time a shuttle returns home, it brings back a portrait of what the planet looked like that particular week, often revealing further evidence of a deteriorating world: great burning sections of Amazonian or Indonesian jungles with palls of smoke spread out over hundreds of thousands of square miles, deltas choked with soil because the forests that once held that soil are gone, rivers of black soot pouring from burning Kuwaiti oil fields. Each returning mission unfolds the ongoing folly of the human race in a kind of time-lapse photography. One mission might return with a picture of the thin line of a road six hundred miles long running through the heart of a canopied forest. A few missions later pictures of the same area reveal a series of smaller roads extending like ribs off the spine of the main highway. Later, houses and villages pop up while the forest disappears around them. Other pictures reveal that large bodies of water in arid and semiarid sectors of Earth, like the Aral Sea in Kazakhstan, Lake Chad in central Africa, and the Colorado River in the United States, are shrinking, being drained by drought or the growing populations around them. Destruction such as the receding shoreline of a lake or a darkened patch of sky takes on more meaning when we see it from space because we can see that the damage is not isolated; it is interwoven and spread by the very systems that create and drive the biosphere.

Of all the ravaged places on Earth, among the worst is a scorched area in Eastern Europe known as the "black triangle," a polluted corridor that runs from northern Czechoslovakia to eastern Germany and then to the Silesian district of southern Poland. Here the engines of the Industrial Revolution have been churning for decades, fueled by the great seams of soft sulfurous coal that lie beneath the surrounding mountains and valleys.

Each day tons of black soot and a lethal cocktail of carcino-

Above: Smoke plumes from burning Kuwaiti oil-well fires in the aftermath of Iraqi occupation in May 1991. Wells north of the Bay of Kuwait and south of Kuwait City burned out of control for nearly a year.

Opposite: Lagoa Dos Patos lagoon in Brazil. Sediment mixed with raw sewage flows from the city of Pôrto Alegre.

Barrett Rock of the University of New Hampshire with graduate student Jan Greczynski in the forests of northern Czechoslovakia. Dr. Rock uses Landsat images as "macroscopes" to assess acid-rain forest damage on a large scale.

Opposite top: Landsat map of the "Black Triangle" region, September 1985. Heavy forest damage and unhealed clear-cut areas appear in orange. A diagonal line just above this orange area is the boundary between the Czecho/Slovak Federal Republic and the former German Democratic Republic. This boundary is also the ridge line along the top of the mountains separating the two countries. Notice that the damage is not as extensive on the lee side of the mountains.

Bottom: Landsat change detection map of a small area of the Black Triangle, taken in 1990. Red indicates areas where trees have died or been clear-cut due to poor health. Yellow indicates areas that have suffered significant increases in damage since 1985. Forty-four percent of these forests have deteriorated since 1985.

gens pour into the sky. The air in some towns is so clogged that lung disease has drastically reduced life expectancy. In eastern Germany eczema and bronchitis reportedly affect half of the children growing up in industrial areas; half of Czechoslovakia's drinking water fails to meet the country's own health standards, and a little farther south in Budapest, Hungary, the air is often so bad that the citizens can't make out their counterparts across the famed Danube. Some citizens seek respite in places called inhalatoriums where they can breathe clean, steamed air for a few minutes.

Not only is Eastern Europe choking on its own industrial waste, but it has become the acid-rain dumping ground for Western Europe. Prevailing winds carry pollutants from as far away as Britain. Across Europe a million acres of woodlands have been damaged—the trees and the ground they grow on are being eaten away. These are the magical forests that once carpeted the continent in the Middle Ages, and that were depicted in the fairy tales "Little Red Riding Hood" and "Hansel and Gretel." Their destruction points up a fact clearly seen from high above the planet: The winds of Earth comprehend no borders; to them it is all the same whether they carry good or ill.

Recently, because of the sweeping political changes in Eastern European nations, ecologists and scientists in Czechoslovakia, Poland, and Hungary have made contact with their Western counterparts hoping to bring the environmental damage in their nations under control. One of these Western scientists is Barrett Rock, an ebullient American botanist who works at the University of New Hampshire. Rock has an unusual background for a botanist. He has not only worked a great deal on studying the cellular structure of plants, but has also spent several years at NASA's Jet Propulsion Laboratory studying the possibility of life on other planets, experience that makes him aware of both the largest and smallest perspectives in biology. In order to effectively study the destruction of forests by acid rain, Rock knew he would need something that he came to call a *macroscope,* a tool that would allow him to look at the problem on a planetary scale, yet still be able to show the microscopic evidence inside a tree's cells that could provide unseen clues to its health. He found this unique tool in the Landsat series of satellites.

Landsats were the first Earth resource satellites. Every sixteen days since 1972 they have constructed a new map of Earth. Unlike

Above: Smokestacks at an electric power generating plant in Czechoslovakia belch plumes of sulfurous, yellow smoke. The acid that results from this pollution contributes to the destruction of forests and wildlife in the area.

Opposite top: Remnants of a pine forest in the Sudety Mountains of Poland where trees looked perfectly healthy as recently as 1978.

Bottom: Blue spruce was planted in the Krušnéhory Mountains because it was believed that the waxy coating on the needles would protect them from pollutants, but in this photo, only the year's new needles appear to be healthy.

the analog cameras that astronauts use, they record the reflected sunlight of everything they see in several electromagnetic wavelengths, or, as Rock puts it, they taste "several flavors of ice cream." Landsats can "see" land, oceans, trees, and so on in ways that human eyes cannot, much like the probes that we have launched to investigate other worlds.

Just as Rock had hoped, one Landsat instrument, the Thematic Mapper, can register the very same damage that a microscope can see in a single pine tree needle, except from 425 miles above the Earth. Rock has carefully compared "the ground truth" he sees under a microscope against the damage that the satellite senses from space. A computer then renders these views in a series of artificial colors that indicate the relative health of the trees. The color green represents healthy trees, which absorb the most light. The chloroplasts within their needles that are vital to photosynthesis are a dense green. But unhealthy needles are paler, not to human eyes, but to the satellite's eyes, and this digital information is represented as brown or yellow. Red represents the trees that reflect the most light and are the unhealthiest, with needles whose chloroplasts are nearly dead.

Rock perfected this method for studying forests damaged by acid rain in New England and then brought it to the forests of Eastern Europe as part of a United Nations study to assess environmental damage in the black triangle. The destruction he found was appalling. To illustrate what he's seen, he often uses a photograph of the remains of a forest on a mountaintop in Poland. Acid rain has turned what were once lush woods into a field of desiccated stumps. The remnants of the trees look as if they have been blowtorched, and the groundwater is ten thousand times as acidic as normal water.

In 1978 no one suspected a thing was wrong with the tall pine in this forest—every tree looked healthy. But by 1987 the ecosystem had begun to collapse. Needles browned and thinned, and the trees grew bare. Today nothing is alive except a few hedges that specialize in growing in acidic bogs.

This is happening throughout the forests of the black triangle. Wildlife today is all but gone; the soil can no longer hold water and is itself often washed away when heavy rains come. It took fifteen thousand years for these woodlands to evolve, but in a matter of a few years they were all wiped out. Eastern Europe's lumber industry lost millions of dollars when the forests died, and

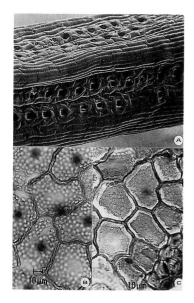

Top: A Scanning Electron Micrograph of a red spruce needle showing epidermal cells.
Below left: A spruce needle in cross section showing healthy chloroplasts.
Below right: A spruce needle in cross section showing severely damaged chloroplasts. The contents of the cell have pulled away from the cell wall.

the few foresters who are trying to replant are being overwhelmed by the continuing destruction.

Rock admits that he has severe doubts about whether the forests of the black triangle can be reclaimed. Human survival in Czechoslovakia and Poland still depends on the burning of soft coal. But he wonders: Are people really surviving? Maybe, he says, the trees are trying to warn us, like the canary in a mine shaft, which dies in the presence of toxic gas. When the canary stops singing, you know you are in trouble. Landsats, he believes, can help us better understand the forests' message. Using them we can sense from high above the planet what is happening inside the cells of a spruce—of millions of spruce. They join the world of the very small with the world of the large and record patterns and problems on scales we couldn't possibly have recorded or understood before. They can tell us when the forests may stop singing, a signal that our own lives may also be in jeopardy.

Only a century ago, Alfred Russel Wallace, who independently formulated the theory of evolution at the same time Charles Darwin did, visited the Amazon jungle and wrote: "The gloom and the solemn silence combine to produce a sense of the past, the primeval—almost of the infinite. The tropical forest is a world in which man seems an intruder, and where he feels overwhelmed by the contemplation of the ever acting forces, which, from the simple elements of the atmosphere, build up a great mass of vegetation which overshadows, and almost seems to oppress the Earth."

It is easy to see how Wallace could have felt this way. One hundred years ago the Amazon rain forest was an unfathomable world unto itself. Traversing it would have been like hacking your way through solid jungle from Cleveland to Los Angeles. It has been growing and evolving for one hundred million years, and the results are stunning.

Biologists estimate that the Amazon basin may house one-tenth of all plant and animal species on the planet. The heart of the forest is the Amazon River, the second longest in the world, which drains an area so vast and wet that it provides 20 percent of all fresh water that pours into Earth's oceans.

In the early days of the space program, the Mercury and Gemini astronauts had trouble photographing the Amazing rain forest from orbit—it was perpetually enshrouded in clouds be-

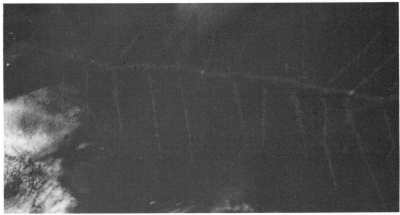

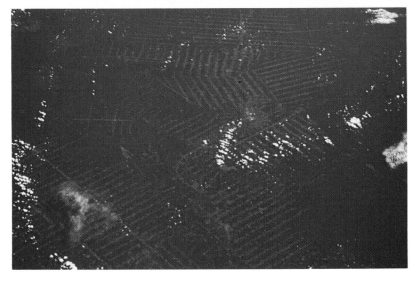

Top: Between 1980 and 1990 approximately 210,000 kilometers of the Amazonian rain forest were cut and burned. The trans-Amazonian Highway is the road through the forest that leads prospective settlers to available land.

Center: In a middle stage of development organized forest clearing penetrates deeper into the interior of Amazonia. Small-scale clearing for ranches and farms follows along the axis of the primary highway and secondary extensions.

Bottom: What began as a single dirt road through the Amazon has become a network of roads and highways.

cause the forest directly recycles half of the rain that enables it to grow so lush in the first place. During shuttle flights in the 1980s, however, the sky over the Amazon developed holes. In some places the clouds had completely vanished because vast tracts of the forest below had been cleared: The cycle had been broken. These holes in the clouds and in the forest below began to appear after Brazil instituted an aggressive slash-and-burn expansion policy that eliminated approximately 75,625 square miles of the Amazon between 1980 and 1990. In 1987 alone, an area the size of Austria was incinerated. The Amazon is not the only rain forest in trouble, but its rapid unraveling has become a symbol of the damage we are doing everywhere as we gobble up the planet's resources.

Since the explosion of the human population began ten thousand years ago, 55 percent of all the world's rain forests have been destroyed, the vast majority in this century. Today we are subtracting the number of species on the planet at a rate that will reduce diversity to the same level it stood at 65 million years ago after a comet or asteroid struck the Earth and wiped out countless forms of life, including the dinosaurs.

Our visits to other worlds and the new view of our own planet that the Space Age has given us have revealed that life is amazingly rare. As far as we know, no other planet in the solar system has defied the law of entropy and come alive. Before we began to explore space thirty-five years ago, some imagined that Venus was a tropical planet with lush rain forests, that Mars might be studded with hardy plants and algae, even that the moon once had oceans. But subsequent visits to all of these places have revealed that they are bereft of life, and probably always were. By comparison, they have shown us how unique and precious life on our planet is—a revelation that is helping us understand the importance of preserving it.

Many scientists believe that the mass extinction that the Earth suffered at the end of the Cretaceous period 65 million years ago followed the impact of a giant meteor. Debris from the collision may have blocked the sun and drastically cooled a planet that was largely tropical. Destruction of the world's rain forests and the diverse life within them represents death on a scale similar to the "great dying" that ended the Cretaceous.

Every microorganism, animal, and plant on Earth contains somewhere between 1 million and 10 million bits of information in its genes. This is the secret of success, the code for outmaneuvering extinction and reversing entropy. Taken together, the genes of all species amount to a global insight into what it takes to survive. E. O. Wilson, the founder of sociobiology, says that we have barely begun to tap this bottomless source of knowledge. But what little we have discovered has already proven tremendously valuable. Consider, for example, the rosy periwinkle, *Ca-*

tharanthus roseus, a tiny plant that originated in the jungles of Madagascar and yields two alkaloids that are highly effective against Hodgkin's disease and acute lymphocytic leukemia. These chemicals, created by the internal alchemy of the plant, generate more than $100 million a year. Less known but possibly just as valuable is the babassu palm, which grows in the Amazon basin. Some day it could become an important source of fiber and oil, if the world will only put it to good use. Five hundred trees can produce 125 barrels of oil simply by doing what comes naturally. And there is the winged bean that grows in New Guinea, an edible plant that rises to fifteen feet in a few weeks and is as nutritious as the soybean—superior from root to tip to almost every other crop cultivated.

Despite this tremendous potential, says Wilson, life is disappearing ten thousand times faster today that it did before the human race existed. We are essentially inflicting irreparable, planetary amnesia, wiping out forever all of these success stories and weakening all life on the planet in the process.

Earth's rain forests contain so much diversity that scientists haven't even been able to catalogue a significant portion of the life in them, let alone all of their genetic combinations. So far 1.4 million living species have been recorded on Earth, and our best estimates of the numbers of unknown species run from 2.6 to as high as 28.6 million. As chain saws and fire sweep through the jungles, no one can say exactly what is being destroyed, only that the destruction seems pointless.

Ironically, the elimination of all of this precious living information leaves behind land that isn't even very productive for the humans who take up its use. The assumption when Project Amazon began was that any soil that could support such lush forests would make prime farmland, but the opposite has turned out to be true. Rain forests are so alive and operate so efficiently that anything that dies or falls from the canopy is rapidly decomposed by the riot of insect and microbial life below and sucked back into the forest's ecosystem. When one walks through a rain forest there are no mounds of rotting leaves because nothing remains dead that long. In the forest, life rules.

When settlers first moved into the Amazon rain forest, mowed it down, and began to plant their crops or graze their cattle, there were no problems so long as the soil remained fertile from the cremated remnants of the forest. But crops rapidly use

Earth view from lunar orbit, one of the most important images of the century. We had to travel beyond Earth before we could see exactly how unique and beautiful our planet is.

up the nutrients left by the incinerated species that once thrived there. In a few years all that remains is sterile, red clay. The lush forest is reduced to a place hardly more arable than a stretch of the Gobi desert—the vast genetic record of the life that once flourished there has been ingested by the crops or cattle that briefly lived on the land, and then moved on.

Recently the Brazilian government has appeared to realize the senselessness in scraping aside large tracts of its forests. The attention that the Space Age has drawn to the destruction of the Amazon has placed growing international pressure on government leaders. Brazil is now attempting to slow the march into the jungle. Its new civilian government recently adopted a revamped constitution with a strong environmental clause, created an environmental protection agency, removed some incentives to develop the forest, and placed new restrictions on burning. The simple passage of new laws, however, will not instantly stem the migration of Brazil's poor into an area that has been promoted as the promised land. And in a jungle the size of the Amazon, enforcement is problematic. Nevertheless, the country's space agency INPE (the Brazilian Institute for Space Research) has taken on the job.

Fittingly, the primary weapons in the battle are satellites. This space-age approach to conservation is the brainchild of INPE's director Alberto Setzer and his staff. To monitor destruction and curb illegal burning, the agency has enlisted the aid of two NASA meteorological satellites to locate areas that are under ecological assault. Like orbiting sentinels, each passes over the rain forest every day—one in the morning, one in the afternoon—faithfully recording every detail below. Images are beamed to Earth and downloaded to personal computers at INPE. There the fires, which seem unremitting, are plotted on a map and the information is telexed to stations throughout the country run by Brazil's new environmental protection agency (IBAMA) where helicopter crews are dispatched to the sites. Hundred of times in the past three years, stunned workers and peasants who are illegally burning remote parts of the forest have been surprised to find a helicopter unloading federal police to arrest or fine them for destroying forest.

The system works extraordinarily well. Usually no more than six hours pass between the time the satellite records a fire and a crew has arrived at the fire's location. In 1989, 5,290 fires were

A comparison of the size of the ozone hole on October 1 from 1987 to 1991 recorded by the *Nimbus* 7 satellite. The ozone hole is pink, and purple represents extremely low ozone levels.

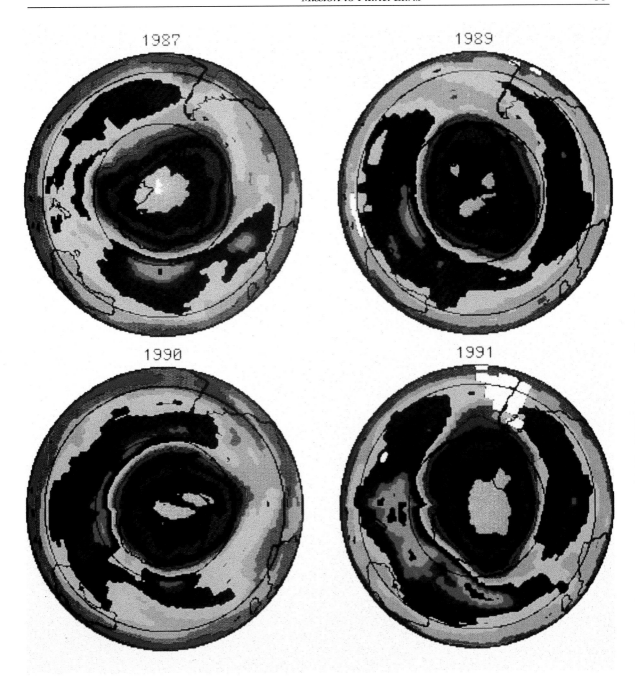

reported, 98.6 percent of them detected by the satellite system. Without the satellites, most of those fires would have gone unnoticed and unstopped. INPE says that between 1989 and 1990 this program has helped reduce the rate of deforestation by 30 percent —saving some four thousand square miles of genetic diversity.

Brazil is currently building its own satellites to gain more control over the information gathered on the forest in the future. Beginning in 1993 the government will launch the first of four of its own satellites. With remote sensing instruments custom designed and assembled by INPE, the satellites will be hoisted into orbit from China using Chinese Long March rockets. Another purpose of this homegrown satellite program is to further Brazil's own understanding of what effects the forest's destruction might have on the global climate. (Studies indicate that it is already having regional effects on rainfall and weather.) Atmospheric scientists reckon that this massive incineration is adding an additional one billion tons of carbon dioxide to the atmosphere each year—about 17 percent of the total that human beings are directly responsible for, which means we are not only sending genetic diversity up in smoke, but worsening global warming and contributing to the destruction of ecosystems elsewhere on the planet. Under these circumstances, carbon, the stuff of life, could soon was just the sort of unpredictable behavior one might expect from clouds.

Ramanathan doesn't believe the formation of this umbrella means that the planet is immune to warming, that it will continuously adjust the global thermostat, like an uncomfortable guest in a hotel room. If global warming raises temperatures enough, he suspects that the cloud buffers could eventually overload. The Earth's oceans might *all* edge toward a permanent El Niño, and the typhoons and droughts might rise up and lash out unabated around the globe.

Nor does Ramanathan believe that his El Niño study provides any pat answers to the future behavior of a warmer Earth, even though it does reveal the complexity of Earth's living systems, the extent of our ignorance, and the shortcomings of the supercomputer models that are used to illuminate the planet's mysterious behavior. This early in our investigations of global processes any conclusive answers are likely to be rare. But the process has to begin somewhere, and only an ongoing effort to refine what we know will enable scientists to create a richer, more accurate pic-

WCRP/ISCCP TEMPERATURE ANALYSIS

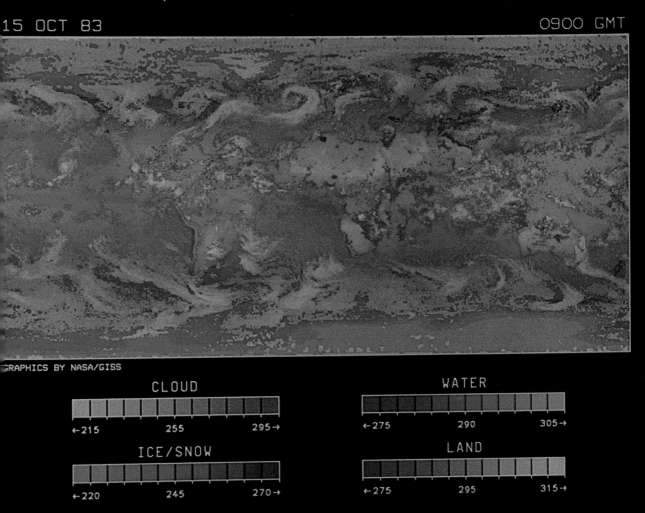

GRAPHICS BY NASA/GISS

CLOUD

←215   255   295→

WATER

←275   290   305→

ICE/SNOW

←220   245   270→

LAND

←275   295   315→

World cloud cover. Scientists like Veerabhadran Ramanathan at the Scripps Institution of Oceanography believe that clouds are among the keys to understanding global warming.

ture of Earth, a struggle that began with the first mapmaker in Babylon so long ago. That's why, where satellites are concerned, Ramanathan is now a believer. For him they provide at least a chance of unmasking the complexities of planet Earth by extending our human senses on a global scale. Without them we could not hope to truly observe the whole world-shaping apparatus and learn its workings, to see how it is handling the pressures we are placing upon it, or even learn what the pressures are.

Goddard, Oberth, and Tsiolkovsky, the old pioneers who looked so longingly toward the stars, would undoubtedly have been surprised to see that their passion for the sky brought forth such unexpected insights about the one world that they did not foresee their ships exploring: the Earth.

Test Tube Worlds: Add Thirty Billion Bytes and Stir

Solving the greenhouse mystery requires an investigation of everything that makes the Earth work—its oceans, landmasses, biogeochemical cycles, and cloud cover—not a simple business. If Earth's mean temperature rises, will the ocean currents change their behavior and rearrange global weather patterns? How much will the ice caps melt? Will massive coastal flooding around the world result? Will cloud cover decrease, or will it increase and reflect more sunlight away from the planet, thereby cooling it? A few clues exist, but nobody really knows. Currently the best chance for finding the answers to these questions, short of waiting to see how the planet itself will react, is to marry the global information-gathering power of Earth satellites with the impressive number-crunching ability of a supercomputer.

Supercomputers transform numbers, turning them into raw material that can be manipulated and interpreted in countless ways. The climate models they build take the digital information recorded by various satellites, combine them with equations, assumptions, and theories, and turn the results into a picture of Earth's behavior. The dynamism of the planet— literally its liveliness—is so complicated that this procedure taxes even computers that can manipulate billions of bits of

Some studies show that since 1880 the average surface temperature of the Earth has risen one full degree. Half of that increase has come since 1950 when industrial activity began to boom at the end of World War II. If the projections of some computer models hold true, the global temperature could rise from 3 to 9 degrees Fahrenheit over the next fifty years—double that which has occurred since the last ice age ten thousand years ago.

SCENARIO B

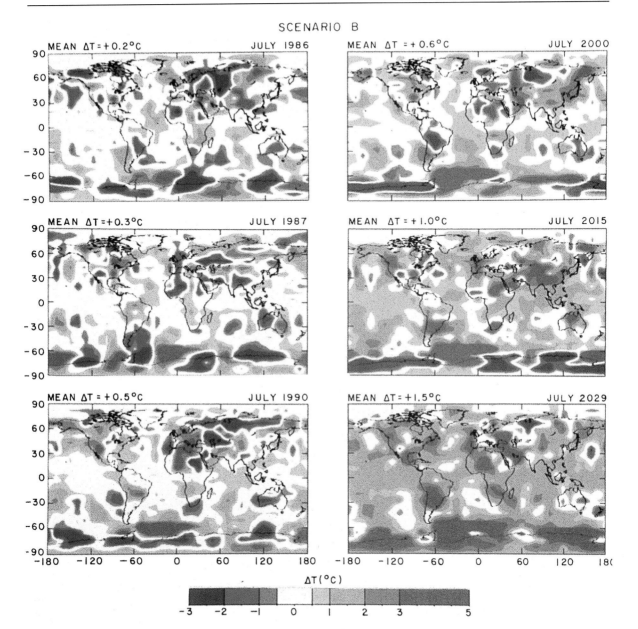

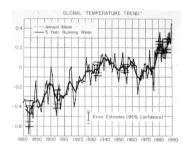

information per second. The result is, admittedly, a timid and halting "virtual" world, but, given a choice, it seems better to experiment with a hypothetical world than the real one.

These models are the modern descendants of the old Earth maps, a manner of synthesizing the planet and finding a way among its mysteries, except that a supercomputer's Earth is an atlas in *three* dimensions. It is Earth in a test tube.

At best, supercomputers merely provide the outlines of a future Earth laden with terra incognita that is vast and impenetrable. By processing more and more data with increasingly powerful computers, scientists are glimpsing detailed features here and there, testing hypotheses, calibrating satellite pictures with information from the ground, comparing the virtual realities of the computer against the hard facts of the real world, and slowly making progress.

The satellite has proven to be so effective as a scientific instrument that scientists, engineers, and governments are now planning to use them to place Earth under even closer, robotic scrutiny in hopes of radically advancing our understanding of what lies ahead. Even in the early days of the V-2 sounding rockets at White Sands, scientists understood that a view of Earth from space might help get to the bottom of certain basic questions they had about geophysics and atmospheric chemistry. But in the mid-1970s, when the full impact of seeing the whole Earth from space had had time to sink in, scientists began to entertain the idea that many orbiting instruments working together might help unravel the intricate machinations of the planet.

In those days NASA developed a project called System Z, an orbiting palette of Earth-observing instruments that would scrutinize the planet's weather, temperature, oceans, atmosphere, and other features to help scientists sketch out a global portrait. The idea gained momentum when a separate Earth-observing platform was incorporated into the Space Station Freedom project. By the mid-1980s recommendations from the Ride Report pushed the concept to center stage. We have to assess the environmental damage we are inflicting on the planet, the report said, so that informed recommendations for governmental action can be made. Later, NASA unveiled a plan called EOS (the Earth Observing System), which was to be part of a larger international program of

This graph *(above)* prepared by James Hansen, director of the Goddard Institute for Space Studies in Manhattan, plots global temperature rise since 1880.

Instruments like the Earth
Resources Satellite (ERS) can view
the planet in many different
wavelengths. Earth satellites are one
of the great unexpected twists of the
Space Age. They have explored our
home planet more deeply than
other space probes have
investigated all of the other worlds
in the solar system combined.

orbiting scientific probes that later came to be known as Mission to Planet Earth.

EOS has been knocked around considerably within the scientific and political communities, particularly with respect to the way the program should be carried out. Should a few large platforms with many instruments be launched, or should many smaller satellites be launched separately? The advantage of large platforms is that they can fly many instruments together and therefore monitor different aspects of Earth's processes as they unfold from precisely the same location and at precisely the same time, an approach that helps scientists better understand the connection between concurrent natural events. They create, in essence, a more unified picture. The disadvantage is that if the launch fails, all of the instruments would be lost. Recently the debate was resolved when several studies concluded that smaller satellites could be launched separately but then flown in close orbital formation, an approach that would accomplish the same simultaneous observations that a dedicated platform could.

EOS is still in the planning stages, but the first U.S. probes will probably be launched at the end of the 1990s. One plan calls for three satellites, each launched eighteen months apart, to carry six instruments into a polar orbit, a route that provides a more complete view of the planet than a satellite flying around the equator. (Later probes may be launched in a much higher geostationary orbit, which has certain advantages in tracking rapidly changing processes on a global scale.) The first probe would concentrate on processes involved with Earth's energy, radiation, and water; the second on the oceans' connections with lower atmospheric chemistry; and the third on solar radiation and processes on land. New expanded versions of each satellite would be launched five years after the first, and the program would continue for at least fifteen years to guarantee a steady stream of accurate and consistent observations over a long period of time. As part of this effort to gather an immense dossier on the planet, the European Space Agency plans a series of Polar Orbiting Earth observation Missions (POEM), one focusing on weather, ocean, and ice processes, the other on land resources and related atmospheric processes. Japan also plans to orbit two additional platforms.

By any standard Mission to Planet Earth is a monumental undertaking, one that rivals the Apollo moon program or construction of the space station. NASA's EOS probes will transmit as much as twice the information currently housed in the Library of Congress, every month, for fifteen years—the equivalent of about six and a half million books a day. Some instruments will be able to see the interaction of life and geology at resolutions just slightly greater than a baseball diamond, from 440 miles away in space.

The fact that politicians and scientists from nations around the world would even seriously contemplate spending as much as $50 billion for such a long-term undertaking says something about the desire to understand our own planet and the urgency we feel about the effect we are having upon it. This awareness exists for the simple reason that we live in the Space Age. In 1968 we departed the planet and, in a serendipitous gaze backward, saw Earth for what it is—a miracle in the middle of nowhere—and it revealed how precarious our future actually is.

Our innate curiosity has unexpectedly led us into a competition between the contradictory aspects of our nature, and we find we must weigh what makes us aggressive, fearful, greedy, and small-minded against what enables us to dream new futures and long to know what makes the universe tick.

What lies ahead in our relationship with Earth? Many believe that we must act quickly if we intend to save the planet. But the truth is, Earth's existence isn't at stake, ours is. We need a vibrant Earth far more than it needs us. Its very diversity is what allowed humans to evolve, and if we continue to damage it, it will not be the planet that we obliterate, it will be ourselves. Ninety-nine percent of the species that have lived on Earth are now gone; we may simply pass into oblivion just as they have. Once we are gone, perhaps, the Earth will take a stab at a new, "intelligent" form of life—one that turns out to be a little less self-destructive—and the experiment will begin anew.

But with luck our curiosity and desire will win out over our destructive side. As we digest the information from our satellites, computers may draw the ultimate atlas—a three-dimensional map that would dazzle the earliest cartographers—and unmask planet-wide mysteries. Within these new digital worlds, we may be able to try out a global experiment or two with impunity, peer into possible futures, and plan accordingly. If we don't we may not survive, which is why we should never deny the seemingly senseless curiosity that drives us.

Following page: The Global Biosphere. The first composite image of the living matter of the planet created by many satellites shows ocean chlorophyll concentrations and land vegetation patterns. The Space Age has revolutionized the way we see the most mysterious world of all—our own.

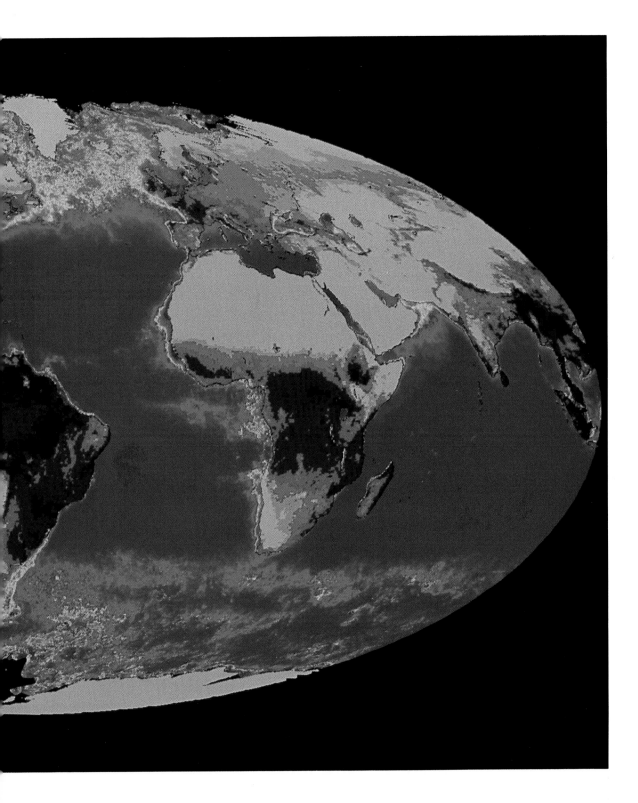

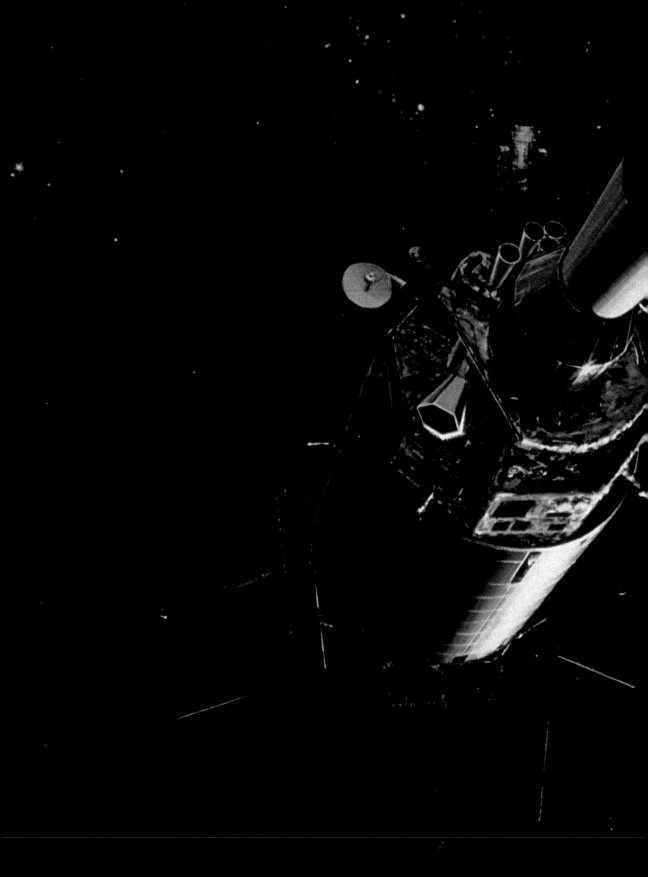

5

*"The fix'd sentinels almost receive
The secret whispers of each other's watch."*

—William Shakespeare

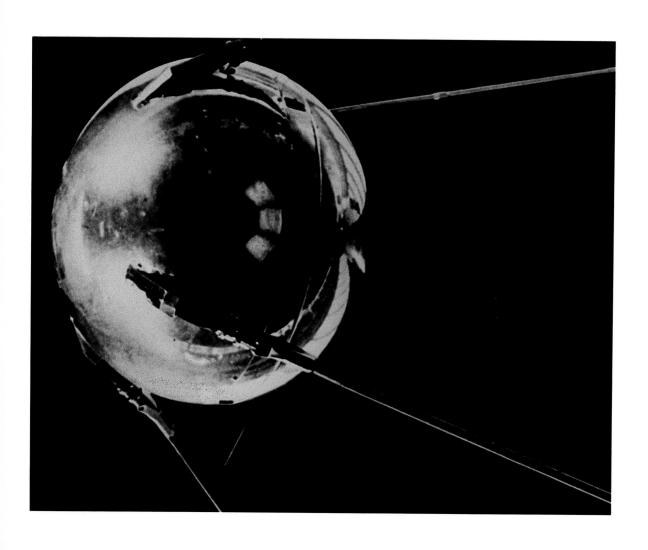

Celestial Sentinels

Above: The Space Devastator. As early as 1939, science fiction was exploring the possibilities of military satellites.

Left: Sputnik, the world's first artificial satellite, was launched on October 4, 1957. It was an aluminum sphere twenty-three inches in diameter, weighed 184 pounds, and was fitted with four radio antennae and two transmitters.

Before the autumn of 1957, nothing circled the immediate vicinity of Earth except asteroids and meteors—ancient incoming artillery from faraway sectors of the solar system. Artificial satellites had not yet been sent into space, and when we gazed into the night sky, it looked much as it had to the ancient Magi who had scrutinized the heavens thousands of years earlier searching for meaning and clues to the future. But in October of that year, the Soviet Union launched *Sputnik,* and the sky changed forever.

At this moment, approximately six thousand man-made objects orbit our planet. About a thousand of these are debris: everything from cameras and wrenches lost by astronauts to discarded rocket boosters. The rest, however, have been placed there on purpose. At night, between May and August, when the northern hemisphere is tilted toward the sun, these objects are visible to the unaided eye as they fly high above the Earth, reflected in the sunlight that strikes their metallic shells. From Earth they may look like nothing more than beads of light, but these mechanical sentinels have revolutionized the way every one of us lives. They have drawn us closer together, accelerated the dispensation of information, and made communication possible where none existed before. In a single generation we have been suddenly enveloped in data created and relayed by communications satellites, weather satellites, remote sensing satellites, navigational satellites, and spy satellites. They come in varieties, like fruit.

A generation ago telephone lines stretched from one wooden pole to another across the landscape of the modern

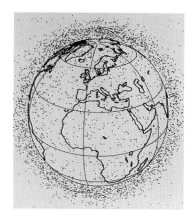

Orbital debris surrounding Earth. Today, six thousand artificial satellites orbit the Earth. About a thousand of these are space junk—everything from cameras and wrenches lost in space to discarded rocket boosters.

world; later, cables were buried in enormous networks. But today microwave signals—digitized words, numbers, music, voices—flood the air, reaching out to geosynchronous communications satellites perched 22,300 miles above Earth, and are relayed almost instantaneously to their intended destinations elsewhere on the planet. In this way the Earth's stores of intelligence and knowledge and experience are transferred and pooled at the blinding speed of light.

Because of satellites, secrets are now harder to keep. Certain military varieties, equipped with cameras, can spy troop movements while others sense the heat of missile launches. Today they can monitor nations' nuclear armories, which has enabled the United States and the republics of the former Soviet Union to drastically reduce their weapons unilaterally and without treaties. During the past thirty-five years these sentinel satellites arguably ensured that the world faced the lesser of two evils—a cold war rather than an all-annihilating nuclear one—and gradually they have encouraged a radical shift in attitudes in Europe, Eastern Europe, and America.

The power of the satellite to communicate what people are experiencing all over the globe boggles the mind. In 1988 nearly half the population of the planet, more than 2 billion people, watched the Olympic Games live from Seoul, Korea. Each evening people can look into the future over much of the world by witnessing satellite images of the weather system that will pass through their regions the following day. In recent years we have watched a failed coup in the Soviet Union, a failed rebellion in China, the rising of the Iron Curtain, and the fall of the Berlin Wall from half a world away. At the flick of a switch hundreds of millions of us witnessed the unfolding events of the Gulf War on television and as we watched via satellite, the military orchestrated it using the same technology. With the help of an enormously complex and secret satellite system, it coordinated the movement of hundreds of thousands of troops, ships, fighters, bombers, and missiles. Allied commanders not only knew the location of many Iraqi troops and units, but also plotted the location of their own —not a simple issue during war. In some places, by setting up nothing more than a portable satellite dish and a laptop computer, soldiers could instantly transform a square of desert into a command center connected to a worldwide information network. Yet, as amazing and revolutionary as these changes have been, it is

Above: Near the Iraqi border in 1991 soldiers of the 101st Airborne check their positioning with a handheld navigation device linked to the Global Positioning Satellite system during the Gulf War.

Left: The *Delta II* lift-off July 3, 1991, carried the eleventh Navstar Global Positioning system satellite into orbit. The Global Positioning system is an American military and civil system for navigation and location. It played a crucial role in the Gulf War, tracking troops, ships, jets, and missiles.

even more astounding that so few people saw them coming.

The satellite that affects us most is the communications satellite, but its genesis was unspectacular. In 1945 a twenty-eight-year-old British officer named Arthur C. Clarke wrote a letter to *Wireless World* magazine suggesting that modified versions of the same German V-2 rockets that were then thundering down on London could be used to launch instruments into orbit high above Earth. These instruments, he said, could broadcast information to much of the planet in a way that would revolutionize communications. Clarke, who would later go on to write the science fiction classics *2001: A Space Odyssey* and *Childhood's End,* had been fascinated with space since his teens. He had already begun to dabble in the art of science fiction after being inspired by *The Conquest of Space,* a seminal book on rocketry written by a cofounder of the American Rocket Society named David Lasser, and by pulp science fiction magazines such as *Amazing Stories* and *Astounding Science Fiction.* But he was more than a simple conjurer of the fantastic. When he wrote the letter mentioning the V-2, he was the chief British officer working on the Allies' Ground Controlled Approach Radar, which he had helped develop with three American scientists, one of them future Nobel Laureate Luis Alvarez.

In his letter, Clarke speculated that in the future "artificial satellites" orbiting at 22,300 miles could remain in a stationary orbit over Earth (just as a horse on a merry-go-round remains stationary in relation to the center pole of the carousel). "Three repeater stations, 120 degrees apart in the correct orbit," he wrote, "could give television and microwave coverage to the entire planet."

No one paid much attention to Clarke's letter, but nine months later he expanded his concept in a full-length article for *Wireless World.* Clarke had boldly entitled the piece "The Future of World Communications," but *Wireless World's* editor toned that down and renamed it "Extra-Terrestrial Relays: Can Rocket Stations Give World-Wide Radio Coverage?" As it has turned out, Clarke's title was the more accurate of the two.

His first paragraph couldn't have been more prophetic:

Although it is possible by a suitable choice of frequencies and routes, to provide telephony circuits between two points or regions of the Earth. . . , long

Opposite top: The U-2 was commissioned by a special presidential committee under President Eisenhower concerned with improving intelligence gathering on Soviet missiles. It was the precursor of the reconnaissance satellite.

Opposite bottom: Picture from a U-2 mission during the Cuban Missile Crisis. U-2s found evidence of Soviet missiles being deployed in Cuba within one hundred miles of U.S. territory. The crisis brought the world to the brink of nuclear war.

MRBM FIELD LAUNCH SITE
SAN CRISTOBAL NO 1
14 OCTOBER 1962

ERECTOR/LAUNCHER EQUIPMENT

ERECTOR/LAUNCHER EQUIPMENT

8 MISSILE TRAILERS

EQUIPMENT

TENT AREAS

CONSTRUCTION

distance communication is greatly hampered by the peculiarities of the ionosphere, and there are even occasions when it may be impossible. A true broadcast service, giving constant field strength at all times over the whole globe would be invaluable, not to say indispensable, in a world society.

Clarke knew that many in the small community of his peers might think his fiction writing was affecting his scientific work, so he pointed out that the Germans had already brought ballistic engineering to the brink of orbital power, which suggested that satellites were inevitable. However, he surmised that they probably would not materialize for another fifty years or so.

In this he was wrong. The first communications satellite was launched by the American military a mere thirteen years later in 1958. The one thing Clarke had failed to foresee was how strategically important would become the satellite for the military.

"It has been incredible what has happened," he said in a recent interview. "And I don't think anybody could have reasonably expected the development [of space] as quickly as it came. The pacing factor was not technology, but politics. You might say the political accident of the Cold War is really what powered our drive into space. If it had been a peaceful world, we might not even have the airplane, let alone landed on the moon." [9]

Following World War II the American military meant to keep the Soviet Union (and later China and their satellite nations) under close watch using recently developed radar, camera surveillance, and other electronic monitoring technologies. Reconnaissance missions were flown in a steady stream along the borders of the Soviet bloc, submarines rigged with the latest high-powered bugging devices eavesdropped off their coastlines, tapped into communications cables, and kept watch on Soviet submarines launching ballistic missiles. The idea was to keep the Soviet Union beneath an electronic shell and draw from it all of the information that this array of technology would permit.

---

[9]   Clarke does not actually lay claim to conceiving the idea of geostationary orbit, insisting that it was perfectly obvious from the time of Kepler and pointing out that Tsiolkovsky "took the concept for granted but did not develop it." Clarke credits an Austrian writing under the name Hermann Noordung with the details of a manned space station and its placement in stationary orbit. Hermann Oberth, in his 1923 book *The Rocket into Planetary Space,* also linked space stations and communication, writing that mirrors reflecting sunlight could be used where cable connections and electric waves might not reach.

But the electronic shell was limited. The Soviet Union was enormous, and reconnaissance flights, land-based radio, and radar could only penetrate so far. Shortly after the war, however, when scientists at White Sands began launching small scientific payloads in the noses of von Braun's rebuilt V-2s, they made a startling discovery. Many of the rockets had been rigged with cameras attached to their tail fins that filmed the Earth receding below as the rocket soared upward. When the scientists looked at these films, they were surprised to find crystal-clear images of the desert and mountains above which the rocket had just risen. They had assumed that the images of Earth below would be as obscured by its atmosphere as the stars were when they looked at them through a telescope. What they hadn't realized is that stars are much farther away, and their images are more easily obscured by Earth's atmosphere.

Von Braun illustrated this simple truth during one of his speeches at the Army War College in the late forties. In a simple bit of show and tell, he held up a piece of tissue paper and said that it represented the atmosphere. It was opaque when you tried to look through it at something far away, but—and then he smoothed the paper flat over a photograph—when you held it close to something, you could see right through it.

The military immediately saw reconnaissance possibilities here. In the late forties and early fifties the RAND Corporation conducted a series of studies for the Air Force on the satellite as an instrument for spying. These published reports culminated in 1954 with a lengthy document entitled "An Analysis of the Potential of an Unconventional Reconnaissance Method." RAND suggested that the Air Force waste no time in developing and launching "an efficient satellite reconnaissance vehicle as a matter of vital strategic interest to the United States." The report also suggested that such a device could transmit information almost instantaneously from half a world away, a tremendously compelling concept as the chill of the cold war deepened.

The Air Force asked Lockheed Missile Systems to develop the hardware for the first spy satellite. Engineers at Lockheed closely followed the first of RAND's recommendations, and designed a satellite outfitted with a television camera that could broadcast recorded images instantaneously to receivers at a base located on friendly territory below. There was a hitch, however. Television images radioed to Earth were far from sharp; the technology sim-

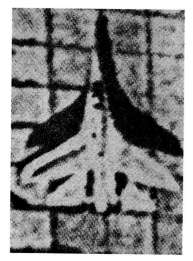

Satellite photos of the Soviet SU-27 Flanker and the MiG 29. These pictures were taken by an American reconnaissance satellite called Big Bird and released inadvertently by the U.S. government.

ply couldn't resolve them with photographic quality.

As an alternative Lockheed also recommended an innovative satellite with a powerful camera that could photograph enemy territory. On a command from Earth the satellite would periodically slow down, lower its altitude, and jettison a cartridge of exposed film. As the cartridge drifted beneath a parachute over the Pacific Ocean, a jet trailing a trapeze pole would hook the falling cartridge out of the sky and bring the precious information home.

Measures like these seemed prudent in the West, because following the war, communism appeared to be making terrifying headway. In 1948 the Soviet Union blockaded East Germany, and the United States retaliated by airlifting supplies to Berlin. In 1949, the same year that NATO was formed, Chiang Kai-shek's Nationalist government fell in China to the Communist forces of Mao Tse-tung. Julius and Ethel Rosenberg were convicted in 1951 for providing atomic secrets to the Soviet Union, and Senator Joseph McCarthy charged that he knew of fifty-seven loyal Communists working right under the Senate's nose in the State Department, although in a subsequent investigation the Senate Foreign Relations Committee couldn't find even one. The paranoia was palpable; events everywhere were cast in hues of red.

By 1957, both the United States and the Soviet Union, driven by the currents of politics and their military needs, were poised for the space race. *Sputnik* was not, of course, a spy satellite, but it clearly illustrated a spy satellite's capabilities. By the early 1960s spy satellites had become as essential tools in the conduct of the cold war, and the sky suddenly became filled with them. Again, the events that followed were unexpected.

In 1967 President Lyndon Johnson was in Nashville to address 125 local officials and educators at the governor's mansion. Johnson was supposed to speak on education and poverty, but he had recently come under fire from certain members of Congress for spending too much money on the space program, and rather than addressing poverty that day, he decided to answer his critics.

"I wouldn't want to be quoted on this," he told the small convocation, "but we've spent thirty-five or forty billion dollars on the space program. And if nothing else had come out of it except the knowledge we've gained from space photography, it would be worth ten times what the whole program has cost. Without the

Brilliant Pebbles. Brilliant Pebbles is a strategic defense program for a space-based interceptor. Here, a Brilliant Pebbles satellite fires its rocket motors on its way to intercepting a ballistic missile. Each vehicle carries a single interceptor that collides with an enemy missile, destroying it on impact.

satellites, I'd be operating by guess and by God. But tonight we know how many missiles the enemy has, and it turned out our guesses were way off. We were doing things we didn't need to do. We were building things we didn't need to build. We were harboring fears we didn't need to harbor." That summarized the most unexpected result of satellite spying. Rather than exacerbating cold war tensions, satellites had reduced them.

Today's military satellites monitor the development of new weapons, the construction of new bases, and the movement of troops. The progenitor of this satellite system in the United States was called MIDAS, the Missile Defense Alarm System, which went to work in the early 1960s high above the planet, programmed to stand an unblinking watch. If it spied the multiple blasts of a preemptive nuclear attack, MIDAS was to warn the American military, which, in the insane logic of nuclear war, would launch enough weapons in return to annihilate the Soviet Union. The message was simple: You may get us, but our system assures that we will also get you, and it became the cornerstone of American military policy, aptly known as MAD—Mutually Assured Destruction.

The spy satellite system that supported this MADness revolutionized world politics. Never before had two such powerful enemies known so much about the military plans of the other. The satellites were informants, stool pigeons in the sky, and the knowledge they gathered became a check on aggression. MIDAS removed what had always been the most powerful weapon in any aggressor's arsenal—surprise. In a lethal game replete with deceit, these satellites made deceit almost impossible.

In the United States, the headquarters for this sort of work is a place known as NORAD, the North American Aerospace Defense Command. NORAD watches the world from a hole in the belly of a 100-million-year-old mass of granite in Colorado called Cheyenne Mountain. In the 1960s, when fear of a Soviet nuclear attack was at its height, the Air Force blasted away 690,000 tons of the mountain's interior and filled it with fifteen steel buildings where today a total of eleven hundred people work during each twenty-four-hour period. It is, in fact, an underground city, with its own generators, air purifiers, medical facilities, water, and food supplies—a monument to the cold war. Two thirty-ton blast doors, encased in concrete collars like those of a bank-vault door, separate NORAD from the rest of the world. The offices are encased in

Brilliant Pebble on its way to intercept an enemy missile.

rock a third of a mile thick, which was designed to withstand a direct nuclear blast, but the power of the latest weapons makes survival today highly unlikely. Inside, the cubicles are unspectacular, not unlike the ones you'd find in any standard office building. There are no cavernous Dr. Strangelove war rooms, nothing very dramatic at all. Yet it is linked to a network that enables the Pentagon to keep its satellite eyes and ears focused on every spacecraft in the sky and on a good deal of what is happening on the ground.

Four types of satellites gather NORAD's intelligence. One variety, originally code-named LaCrosse, beams microwaves and can resolve, through clouds, objects no more than three feet across, augmenting the work of a corps of optical satellites when bad weather blocks their views. The optical variety of satellite is far more powerful than its film-dropping predecessor; new semiconductor technology enables them not only to see with photographic quality but also to transmit their images instantaneously. The military expects that the latest of these satellites will resolve objects that are as small as five inches across!

"Ferret" or SIGINT satellites are a third kind. Ferrets listen; their extremely sensitive receivers can monitor radio and microwave transmissions from thousands of miles above Earth. They are able to eavesdrop on radio conversations, and can even capture signals that determine the trajectories of missiles. They are the workhorses of the spy satellite system. An early ferret called Rhyolite operated through the 1970s and could snoop out signals that aimed Soviet and Chinese missiles and at the same time monitor as many as eleven thousand conversations going on between telephones and walkie-talkies.

Finally, to move this information around the planet and down to receiving stations on Earth, the planet is ringed with a fourth variety of orbiter, military relay satellites, which trace their roots to the satellites that Arthur C. Clarke wrote about in *Wireless World* forty-five years ago.

NORAD also keeps a watchful eye on what is already in orbit around Earth—since Sputnik, 18,300 different objects have been sighted. When a Soviet, French, Chinese, or Japanese rocket departs Earth, NORAD spots it, tracks it, analyzes it, and relays the information instantaneously to the air-men and -women working its computer terminals. It can sense the heat of a rocket's exhaust trail, create an image, and transmit it instantly to NORAD. The system makes about thirty-five thousand satellite observations a

The TDRS just prior to deployment. The TDRS system provides almost constant communication between satellites, space shuttles, and Earth. It can track twenty-six satellites at a time and relay enough information to fill a twenty-volume set of encyclopedias in one second.

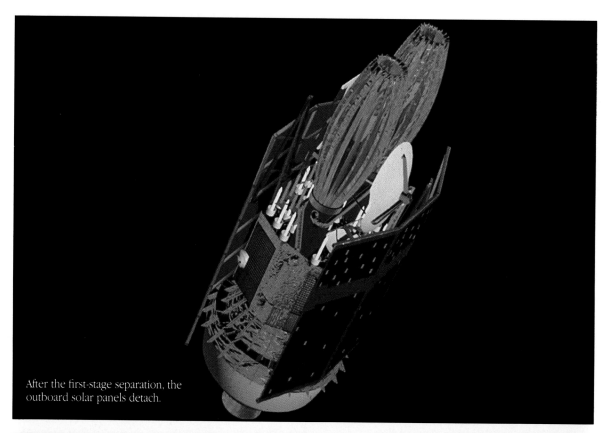

After the first-stage separation, the
outboard solar panels detach.

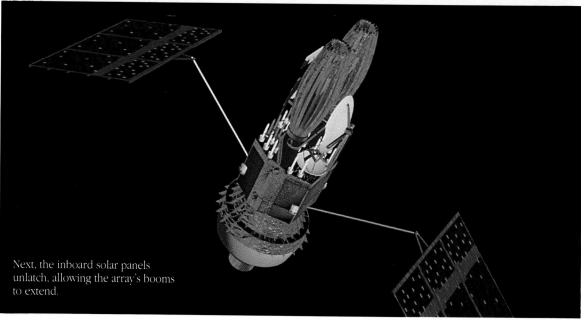

Next, the inboard solar panels
unlatch, allowing the array's booms
to extend.

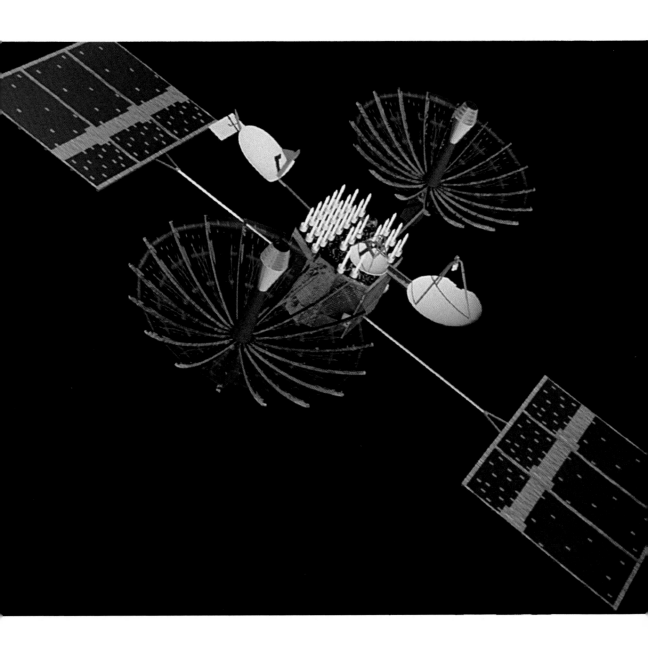

Fully deployed, the TDRS goes to
work.

day, everything from weather satellites to discarded boosters to Soviet reconnaissance cameras. It even knows, at all times, the location of a hand-held camera that was lost a few years back by a shuttle astronaut.

NORAD can track all of these objects because, among other things, it communicates with telescopes that can photograph, in detail, an object the size of a basketball orbiting twenty-two thousand miles above the planet. In April 1981, following the first shuttle launch, these telescopes determined the amount of damage sustained by the heat tiles of *Columbia* before it returned to Earth. It found there was some minor damage, but nothing serious, and the shuttle landed without incident. The cost of operating these global sensory organs is an estimated $5 billion a year, and the entire system is valued at more than $50 billion.

Military satellites have put old enemies like the United States and Soviet Union on intimate terms; they turned simply wary adversaries into unusually well-informed adversaries, and remain crucial even as the cold war fades. Today, knowing if anyone is breaking treaties is more important than ever, particularly as the various Soviet Republics struggle to keep the vast arsenal it has distributed throughout Eurasia under some kind of central control. Undoubtedly both American and Soviet satellites are keeping their electronic eyes on breakaway republics where arms might fall into the hands of weapons merchants or suddenly be used to settle any of the regional and ethnic conflicts that might arise. Furthermore, the number of nations capable of building a nuclear arsenal is rising. North Korea is close to the development of nuclear armaments, Israel already has them, Iraq was closer than many analysts thought (satellites are clearly not infallible), and Iran has recently gained important nuclear technology.

Throughout the cold war, military satellites have played this silent, sentinel role, but recently they took on additional tasks when war briefly broke out in the Persian Gulf.

The Defense Support Program (DSP) satellite. Since 1970, North American Aerospace Defense Command (NORAD) has relied on DSP satellites as the space element of their Tactical Warning and Assessment System.

On September 1, 1939, German panzer units supported by waves of Stuka dive bombers cut through the stunned defenses of the Polish Army and overran that country in four days. The Poles fought valiantly but their cavalry and horse-drawn artillery, throwbacks to the nineteenth century, were no match for the coordinated, multifaceted offensive of Germany's modernized army.

The method of attack came to be known as the "blitzkrieg"

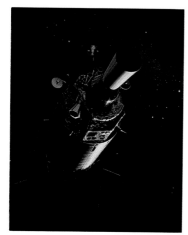

Above: An artist's conception of the Defense Support Program Satellite on its rounds in space.

Right: A DSP satellite in the shuttle cargo bay just before it is deployed.

or lightning war. It fused the power of the internal combustion engine (in aircraft, tanks, and trucks) with radio (to coordinate the attack) into a new kind of power that quickly led to the fall of Norway, Denmark, Belgium, and France. The blitzkrieg changed the face of war, and it was terrifying.

In the eyes of some military analysts, the world glimpsed a new kind of blitzkrieg in the Gulf War. From bases all over Europe and the Middle East, coalition forces launched coordinated attacks that crippled Iraq's vastly inferior command and control capabilities, and paralyzed ground troops and tank divisions. Though not every aspect of the attack ran with absolute precision, it nevertheless unveiled a kind of military power never before seen, something analysts have termed *hyperwar.*

Without satellites hyperwar is impossible. In the Gulf, SIGINT or ferret satellites in geosynchronous orbit tapped Iraqi communications while reconnaissance satellites scanned the same region where the Sumerians had invented writing fifty-five hundred years earlier. Together they plotted the movement and location of Iraqi ground troops whenever and wherever they moved. Under a program called "Constant Source," fighter pilots and foot soldiers alike used small UHF receivers to tap into a global data network. Never before did so many "grunts" have so much information at their fingertips. With a few keystrokes they could pull orders, weather reports, and coordinates on their location literally right out of thin air. This combination of military hardware with an ability to rapidly gather and communicate torrents of information created an entirely new kind of military strength, a power that knew better than ever not only where the enemy was but the depth and location of its own power as well.

"Satellite data and airborne radars have replaced the cavalry scout and the foot patrol as the commander's eyes," military analysts Lt. Colonel Rosanne Bailey and Lt. Colonel Thomas Kearney wrote in *Defense News* shortly after the war. "And although the fog of war was not eliminated, General Schwarzkopf's view of the battlefield exceeded anything before possible. There was far less uncertainty regarding the enemy's vulnerabilities."

With power such as this would Napoleon have won at Waterloo or would Hannibal have succeeded in conquering Rome 250 years before the birth of Christ? If Winston Churchill had had even one satellite during World War I, that war might have ended much sooner. In 1915, as the young head of the Admiralty, he suggested

Arthur C. Clarke *(above)* conceived the geostationary satellite and first published his idea in an article entitled "Extra-Terrestrial Relays: Can Rocket Stations Give World-Wide Radio Coverage?" in *Wireless World* magazine at the end of World War II.

Opposite: Cross section of the living quarters in a moon ship, drawn by Fred Freeman and published in *Collier's* magazine in 1952. Arthur C. Clarke's first vision of a satellite included technicians who would go "round unplugging the burned-out vacuum tubes all the time." The transistor made vacuum tubes obsolete.

that the British Navy run the Turkish blockade at the Dardanelles and take Constantinople at the Bosporus, where Sir Herbert Kitchener's troops would land, and then sweep through the Balkans into Austria from the east. This, he believed, would cut off the Turks, squeeze German and Austrian forces between their eastern and western flanks, and shatter their lines of defense.

The fleet's naval commander did in fact take warships up to the straits but then hesitated, convinced that the fleet would be annihilated running the blockade. In the end Churchill was forced to rescind his order, and he was blamed for the fiasco, which nearly ruined his career. (He remained a political outcast until his suspicions about Hitler proved true twenty years later.) But the plan could have worked, and a single satellite photo would have proven it—enemy defenses were weak. Roger Keyes, a naval commander who supported Churchill's order, steamed through the straits in 1925 and said, "It would have been even easier than I thought. We simply *couldn't* have failed... [B]ecause we didn't try, another million lives were thrown away and the war went on for another three years."

The Gulf War revealed more than the destructive force of satellites; it illustrated how powerful the rapid, global movement of information can be. One satellite is powerful, but many of them working together is like the difference between a single neuron and a brain. Their effect is synergistic; the power of the whole becomes far greater than its parts and it is this growing information network that is changing the world so rapidly today.

In 1956 the first voice cable was laid across the bottom of the Atlantic Ocean between the United States and Scotland, but because it had been designed to handle only thirty-six phone calls at a time, its capacity quickly fell short of demand. There was talk of resurrecting Arthur C. Clarke's crazy idea about communications satellites, but reputable scientists and engineers had serious doubts about whether anyone could build such a contraption, or if they could, whether they would ever be able to get it into space.

During the ten years that had intervened between Clarke's *Wireless World* article and the first transatlantic phone call, his wartime brainstorm was still waiting for both satellite and rocket technology to mature. A central problem was size and reliability. In 1945 not even Clarke had envisioned his communications satellites as the compact digital electronic packages that they are

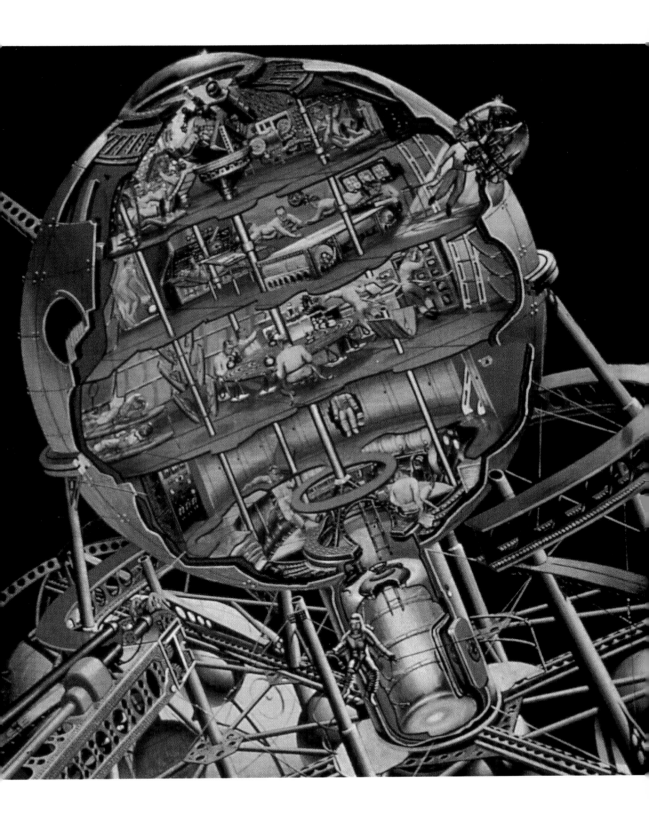

today. He had assumed, as everyone else up to that time had, that the first man-made satellites would be part of larger orbital space stations, complete with an on-board staff of technicians. Clarke recalled. "...I didn't believe, or didn't imagine, that communications satellites could be anything but manned space stations because there would have to be engineers going round unplugging the burned out vacuum tubes all the time." But the shortcomings of vacuum tubes evaporated in 1948, when Bell Laboratory's John Bardeen, Walter Brattain, and William Shockley announced their invention of the transistor. Suddenly the size of electronic equipment shrunk. Transistors were smaller and more reliable than vacuum tubes, and soon more complex equipment was being crammed into tinier areas.

While these advances were being made, von Braun and Korolev were busy designing ICBMs for their respective military bosses, hoping for the opportunity to carry payloads into space. In 1957 Korolev got his chance and launched *Sputnik* up over the broad steppes of Tyuratum. The United States quickly followed with a series of scientific satellites of the Long Playing Rocket type while at the same time the military pushed development of a suite of devices that anticipated the wider use of satellites in the future.

One of these devices was SCORE (Satellite Communications Repeater), the first communications satellite and a wonder of modern technology when it was launched in 1958. Engineers outfitted it with a recorder and designed it to transmit orders to distant locations. As it passed over a base in, say, the United States, a message could be uploaded and recorded. Then as the satellite drifted over the Pacific it would play back the encoded message and transmit it to a ship somewhere between, say, Hawaii and Guam. This process took perhaps half an hour and accomplished what the slow, piecemeal approach of radio relays and cables sometimes took days to manage. SCORE only lasted two weeks before its batteries died, but it illustrated the tremendous potential of satellite communication.

The first civilian communications satellite didn't fly until 1960. Strangely enough it traced its roots to a weather experiment NASA had been working on in 1958, the same year two Bell Laboratory scientists, John R. Pierce and R. W. Kompfner, happened to come across a published picture of a large inflated plastic sphere that NASA was planning to place in orbit one thousand miles high. The balloon was part of an early experiment to measure the den-

sity of the upper atmosphere: the faster its orbit decayed, the more dense the atmosphere.

Pierce realized that with a strong enough transmitter and receiver, a signal could travel from one point on Earth, carom off the balloon, and be received at another point far away. It would be like using a communications tower that was one thousand miles high. NASA agreed to collaborate with Pierce and his Bell Labs team, and on August 12, 1960, a Delta rocket carried *Echo I* into orbit. That same day the first instantaneous satellite message in history was successfully beamed from Goldstone, California, to Bell Laboratories in Crawford Hill, New Jersey, and a revolution in communications was under way.

Despite *Echo's* success, its rudimentary technology would clearly have to be improved. *Echo* had been appropriately named because it was really nothing more than an inflated mirror that required that extremely strong transmitters and outrageously powerful receivers be at both ends of the signal. Today this art of transmission and reception is so refined that signals for global positioning satellites are bounced off micrometeorites streaking to Earth during the few seconds they exist before burning up in the atmosphere. In 1960, however, Pierce estimated that *Echo* reflected back only one millionth of a millionth of a millionth of the energy beamed up to it. So Pierce and his Bell team began work on a much more sophisticated satellite, which they called *Telstar,* an instrument that resembled the mirrored balls that rotated above dance floors in the 1940s and later made a brief comeback in disco clubs during the 1970s. It was essentially an orbiting microwave relay station designed to transmit television signals, two-way phone conversations, and high-speed data across the Atlantic Ocean.

Telstar, like *Echo* before it, attained celebrity status. After it transmitted its first live broadcast from the United States to Europe, a poll showed that two out of every three British citizens recognized the name, and more than half said they actually saw the first broadcast itself. A British group called the Tornados recorded a hit song entitled "Telstar" and it became the top-selling single of the year in both the United States and Britain.

Since *Telstar* and *Echo,* satellite communications have exploded. In 1964 an international consortium of nations founded Intelsat, a nonprofit business venture specializing in international satellite communication. The first of the Intelsats to go up was

Telstar I. Telstar I was the first active communications satellite. It also transmitted the first live television broadcast across the Atlantic. *Telstar* was only thirty-five inches in diameter. It made communications possible for twenty minutes at a time between cities in Europe and the northeastern United States where large steerable antennae had been built.

called *Early Bird,* a dark drum-shaped instrument a little more than four feet high with a small flared cone at one end that could handle 240 phone calls at one time. If you wanted to make room for one television transmission, however, the calls had to be cut off. Today the most advanced international satellites are nearly fifty feet long and can carry 120,000 calls simultaneously. Soon, rather than relying on microwave repeater towers, newer satellites will beam intense signals directly from space to cellular telephones in cars and trains and jets, greatly increasing their range and usefulness.

These advances, like all advances in human communications over the eons, are having a fundamental effect upon the way we live, even upon human evolution and progress, shrinking the world, compressing time, and accelerating the powerful effect of new knowledge. DNA, for example, is a chemical system of communication, the development of which made life itself possible. The next innovation came with the evolution of the human brain: the connectivity of chemical information transmitters—neurons—reached a critical mass which enabled humans to reason and later to speak. With language, an invention of that same reasoning brain, it became possible to share information between many brains, which subsequently led to the invention of writing, mathematics, and the arts—in effect to culture, which has allowed the knowledge and experiences of many to be recorded on paper, film, tape, and computer disks. This information, from the plays of Shakespeare to the insights of Confucius to the latest news, is now passed through time and across ever-lengthening distances via the telegraph, radio, telephone wire, and, finally, the satellite.

The communications satellite business is now a $6 billion industry in the United States alone, but its power is far more than merely economic. It has embraced us in a common envelope of information and experience, something most recently illustrated in the satellite-beamed news reports of CNN and other networks during the Gulf War, a dramatic fulfillment of Marshall McLuhan's 1970s vision of a "global village." Regardless of any individual's view of the war, the instantaneous coverage created a common global experience. It allowed us, in effect, to bilocate: to be in Saudia Arabia as SCUD missiles streaked overhead, and then, an instant later, to be in Israel, where more warheads fell from the sky; to see Saddam Hussein in his command bunker and George Bush in the White House both sitting before their electronic

The Relay communications satellite was designed to test the possibility of transatlantic communications using microwave radio equipment. Low-Earth-orbiting satellites like Telstar and Relay were quickly replaced by satellites in geostationary orbit.

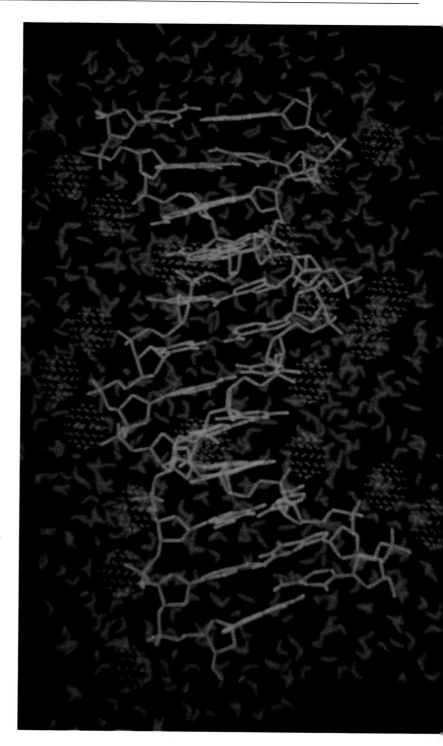

The process that allowed *Homo sapiens* to evolve from *Homo habilis* created changes in the genetic code. The genetic code controls everything from anatomical structure to behavior. In this sequence a computer model illustrates the interaction of DNA molecules with protein.

Right: DNA, the first information and communications system. Fundamental revolutions in communication—DNA, the neuron, and satellites—have shaped and reshaped us.

hearths, drawing different meanings from the exact same images.

Ironically, the satellite is also fragmenting us or, as futurist Alvin Toffler puts it, "demassifying" societies by creating communication specialties for smaller and smaller audiences. Today people in the most out-of-the-way cities can tune in to channels that offer everything from gourmet cooking and science fiction to comedy and the arts. Three million Americans now receive television transmissions directly in their homes via signals scooped out of the sky by a satellite dish on their roofs. The residents of West Virginia joke that the state flower is the satellite dish; they literally line the roads in some rural counties. Dennie Sharp, a kindly grandfather from Mingo, owns one. "You push a button yonder," he says, speaking of the control box in his house, "and . . . you can watch everything that happens in the world. You can watch high waters and big snows. It's something to see."

No longer are viewers restricted to the major networks and the Public Broadcasting System for information and entertainment. Some homes, many in remote sectors of the nation, now have access to 108 channels! A satellite/cable network called Channel One beams educational programs into classrooms, and National College Television targets university students with a special menu of academic lectures and shows. The marriage of communications satellites to cable television systems during the 1980s has further intensified Toffler's demassification, so much so that today nearly 60 percent of all homes in the United States are linked to seven thousand separate cable systems fed by twenty-five different satellites, with the average cable customer now able to tune in twenty-seven different channels.

In 1979 only a small percentage of American homes had linked themselves to cable television systems, partly because the system itself was in its embryonic stages, but also because there was so little to watch. The public wanted variety. But unlike the major networks, cable systems did not have nationwide transmission systems linked by microwave towers or underground links called "land lines." Programming, therefore, had to be created locally, and many operators weren't up to the job or the financial burden that that entailed. This situation changed dramatically when Home Box Office and Turner Broadcasting found that they could beam a local signal to a satellite and thereby reach cable operations nationwide. Suddenly land-line transmissions became unnecessary, or at least less important, and local programming

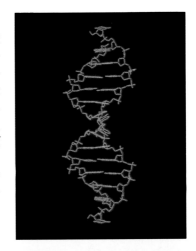

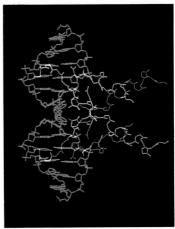

Top: DNA in its natural state in a watery solution. When bound to a specific protein DNA exhibits a "kink" structure, seen in the middle of the molecule.

Bottom: This "kink" facilitates the binding of the protein to DNA. Under specific biochemical conditions DNA will cut the protein at a certain point, possibly eliminating errors in the genetic sequence. In a sense DNA is an ancient precursor of today's global satellite system.

was transformed into national programming. So-called super-stations were born, movie channels proliferated, pay-per-view sporting events materialized, and cable systems were flooded with programming.

By the mid-1980s, this same specialization was evolving in Europe as well. British television viewers who were once provided with one commercial and two government channels—ITV, BBC 1, and BBC 2—may soon have as many as fifteen from which to choose. Already, 700,000 British families have outfitted their homes with satellite dishes for direct broadcast, and before the end of the century that number is expected to include more than half of Britain's homes, generating up to $1.3 billion in advertising. Since 1986 the number of networks in France has risen from three to six; in Italy, Romans now enjoy twenty-five channels; the same trend is evident throughout the rest of Europe. Even Arthur C. Clarke, who has lived to see his satellite brainchild come to fruition, finds the changes astonishing:

> When I grew up in the 1930s everybody who lived in the so-called developed world lived in a very small area, psychologically as well as physically. I was nearly 20 before I left my home town and went to London, all of 150 miles away. I never made any international phone calls because there weren't any. I never saw any pictures of far countries except occasionally when I went to the cinema and there you'd see a flickering black and white picture, travelogues, "as the sun sets slowly in the west," that sort of thing. That was the only glimpse we had of the world. Now a total transformation. Every night everybody, at least in the developed world, sees aspects of the whole human family, all over the world, good and bad. They know they can talk to anyone anywhere. Psychologically, they live in a totally different society.

A spectacular orbital photo of tropical storm Sam in the eastern Indian Ocean. Weather satellites can track storms, follow weather systems, and save thousands of lives.

Developed nations aren't the only sectors of society that the satellite is transforming. It is also drawing the third world into what Clarke calls the "tele-family of man." China, Mexico, India, Indonesia, and parts of Africa are already linked by satellites that specifically service those nations. Argentina is developing satellite technology, and Brazil is working with China to launch several of their own communications and remote sensing devices.

Top: Hurricane Hugo. In 1989 Hurricane Hugo tore through the Virgin Islands, Jamaica, and Puerto Rico and then across the southern United States. It was one of the most powerful storms of this century. Satellites watched it form, allowing meteorologists and local authorities to warn those in its path. Despite its enormous power only forty people died, many from downed power lines after the storm.

Verner Suomi invented the spin camera weather satellite, which made it possible in 1966 to watch global weather systems from space for the first time. Every twenty-four minutes it beamed back a full view of the planet and all of its weather systems. Today's world storm warning system called McIDAS is another of Suomi's innovations.

Satellites make tremendously cost-effective tools for unifying poor countries with large and disparate populations. India first began to engineer a satellite system of its own in the 1970s. One of the largest and most diverse nations in the world, India has nearly one billion citizens who live mostly in remote rural farm areas and speak fifteen hundred different languages and dialects.

The Indian government established a dedicated satellite system in 1983, which now beams national radio news, phone calls, weather reports, and television shows throughout the country. Farmers who still use steel plows pulled by yoked oxen now regularly huddle around village television sets and radios to get the latest information on rainfall and snow-melt so they can more effectively schedule their spring planting.

Indonesia, a country of three thousand islands scattered over 5 million square miles of ocean, from Malaysia to New Guinea, with more than 175 million inhabitants, has also put the satellite to good use. Its system handles telephone communications, television and weather reports, and helps create a sense of unification in a nation fractured by its geography. For many who live among the forested slopes of Java or Sumatra, Celebes or Timor, the satellite has opened up a world they would have found unimaginable ten years ago.

We do not normally think of satellites as time machines, but because of their global perspective they can glimpse the future. Satellite cameras watch whole weather systems forming and moving across oceans and continents, tracking the meanderings of the world's clouds, the fluctuations of its air currents and temperatures, essentially painting a portrait of what the next day will bring.

Until recently we had very little idea of what we might find in the sky when the sun came up. Weather prediction for most of our history has been little more than a crap shoot. We've made arthritic joints and bunions into instruments of meteorological prognostication, and substituted folklore for barometers. Charles Dudley Warner, a friend of Mark Twain, best summarized our frustrations about the weather when he said, "Everybody talks about the weather, but nobody does anything about it." Today we may be no better at changing the weather, but we have improved our ability to forecast it, especially in its most destructive forms.

In 1989 when Hurricane Hugo, one of the most powerful storms of this century, pounded the Virgin Islands, Jamaica, and

NASA ATS III 19Nov67 144327Z 17N

Previous page: The first view of Earth from Suomi's spin-scan camera satellite in 1967. Suomi took these pictures and made a movie loop of them, showing how satellite pictures could be used to forecast weather.

Above: A composite of pictures from the McIDAS satellite system providing a global view of Earth's weather systems.

Opposite: A NOAA vegetation index image for the African continent. Satellite technology allows quick and complete assessment of available resources, and can monitor the possibility of drought for African planters and herders. Given enough warning some crops, animals, and human lives can be moved and saved.

Puerto Rico and then slashed through parts of the southern United States, it did $4 billion in damage, but only 40 people died, many from downed power lines after the storm. The death toll was as low as it was because satellites had seen the future in a vortex of clouds and 135-mile-per-hour winds, and the information was passed on to those in Hugo's path.

This storm-warning system is largely due to the work of meteorologist Verner Suomi at the University of Wisconsin, a man of long experience but eternally boyish enthusiasm. His invention of a weather satellite called the spin-scan camera in 1966 was the first instrument that could watch global weather systems form from space, scanning a small part of Earth from high orbit each time it rotated, and then beaming back a full view of the planet and all of the weather that enveloped it every twenty-four minutes.

"This was the first time we could see the weather in motion," Suomi recalls. "I took the first pictures and made a crude movie loop that went round and round in the projector. You can see where in my excitement I'd pricked my finger and smeared blood on the film. I showed it at a meeting of the American Meteorological Society. I went down into the audience and sat there mesmerized by those pictures."

One of Suomi's students recently used a complex computer system linked to a microwave on a polar-orbiting satellite to predict the winds of Hurricane Gilbert, a monster of a storm that struck the Caribbean and southern United States in 1988. "He [the student] went down to the hurricane center in Florida and showed that with a satellite we can do what we used to do by sending out airplanes [into the storm]." The satellite, of course, was far less risky and its view much more complete.

Suomi says one of the biggest problems with a hurricane is convincing people to get out of the way—they simply don't believe that it's dangerous. Today weather satellites make celebrities of the storms, beaming something like a wanted poster directly to potential victims' television sets.

"Now because people can see . . . what a roaring monster the storm is, they get serious about protecting themselves," says Suomi. "Even though Gilbert flattened places like Jamaica, the number of lives lost were in the hundreds. Thousands of people who could have been killed got out of the way. The people we save, we don't know their names. It would be nicer if we did."

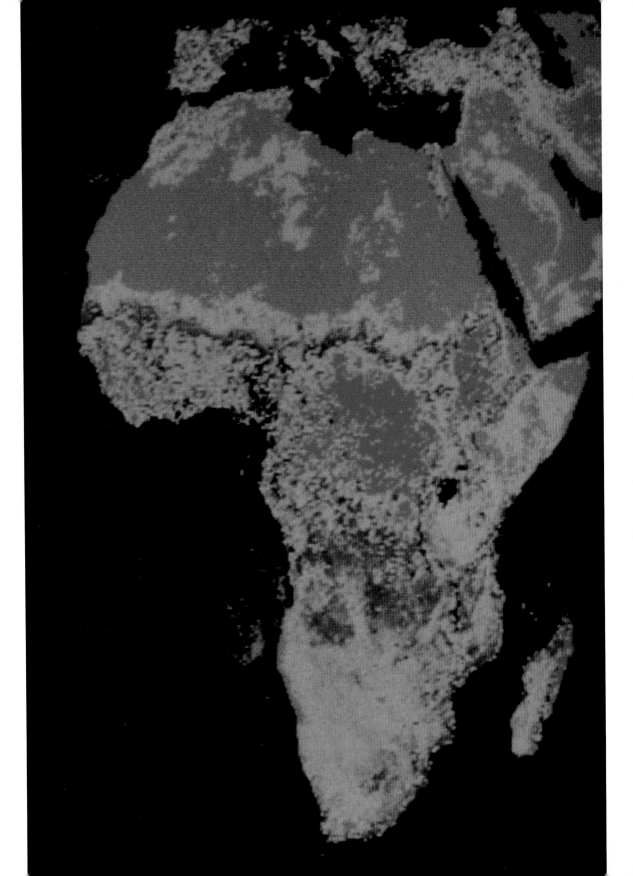

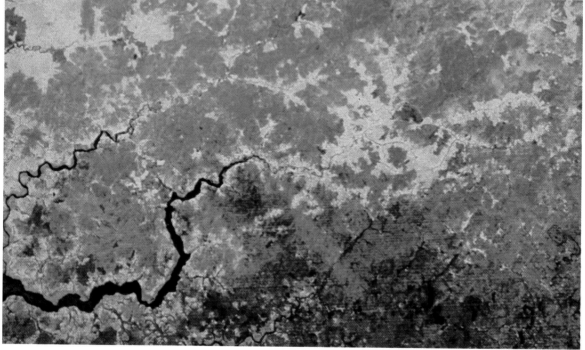

Not all disasters unfold with the fury and speed of a hurricane. Some, like drought, take months to reveal themselves, but can be even more deadly. Satellite receiving stations in places such as the arid outback of the African republic of Niger are now used to monitor rainfall, weather patterns, and crop growth over enormous areas that are constantly threatened by drought. The sight of a satellite dish may look out of place in a region where roads, plumbing, newspapers, and televisions are few and far between, but this very lack of infrastructure is what makes satellites so valuable to the people who live there.

Today they can measure the response of vegetation to weather conditions by downloading information from several polar-orbiting satellites operated by the United States, which are busy scanning the entire planet and its atmosphere, recording the activity of chlorophyll in plant life by sensing certain wavelengths of reflected light. Chlorophyll and water go hand in hand, so where chlorophyll exists in healthy amounts, sufficient water must also exist. The view of these satellites is so expansive that they can measure chlorophyll activity across whole continents, yet still so sensitive that they can discriminate information on a microscopic scale.

Another satellite known as Meteosat is able to measure rainfall over large areas, while a Landsat tracks the type and density of crops and pasture land and helps locate natural watersheds. This information is then broadcast so that herders in Africa can tune in to their radios to find out where these satellites have spotted greener pastures. In the past a Nigerian shepherd might have allowed his cattle to graze an area to dust while there were fresher pasture lands just over the next hill. Local farmers can listen to the same broadcast and learn whether rain is on the way and even get some idea of how heavy it might be.

Given enough warning some crops in marginal regions like Niger can be saved. This same system has even helped predict possible locust plagues by locating the insects' breeding grounds before they spread out and destroy croplands. Whatever the circumstances, the satellites' ability to forecast what may lie ahead allows international relief agencies to mobilize assistance as far as six months in advance so that food can arrive in remote villages soon enough to avoid the famines that often plague certain African nations.

This system is being put to the test right now. The most

Above top: The countryside of Senegal during the wet season.

Above: The same region during the dry season.

Opposite top: Wet season map, Senegal in western Africa. Life in the Sahel region of Africa revolves around water. When the rains come they may only provide enough water to keep the area green for three months. When water is available the farmers and herders must use it wisely in order to produce enough food to last through the long dry period.

Opposite bottom: Dry season map, Senegal. The dry season lasts from October until June, a time when nothing grows and crop production is impossible.

drought-stricken area of Africa, a place known as the Sahel just south of the Sahara that stretches from Niger in the west to Ethiopia in the east, once again suffered poor growing seasons in both 1990 and 1991, placing millions of people in jeopardy. But thanks to the warning system, these nations already know that they may soon face famine. Relief agencies are now poised to provide aid, and if the satellite system forecasts another poor harvest, supplies will be shipped in before food shortages become acute. Even a decade ago, few could have predicted that instruments launched into space could be used in such a variety of ways.

Geo-positioning: New Wave Satellites

Many experts believe that the next satellites that will radically change your life will be geo-positioning satellites. These are the instruments orbiting Earth that enabled soldiers, pilots, and generals in the Gulf War to precisely determine where they and their troops, ships, and artillery were located. Information about the positions of troops that would once have taken days to assemble can now be collected instantaneously.

The United States military has nearly completed the world's most extensive Global Positioning System (GPS). It consists of twenty-four satellites (the last due to be launched in 1993) that will essentially make it possible to accurately locate anything on Earth within twenty-five meters. Impressed with the success of this geo-positioning system, the military is now busy equipping everything from battleships to jets and cruise missiles with GPS receivers and transmitters.

The Global Positioning industry is growing so fast it even has its own magazine *(GPS World)*. Using mapping information from the National Oceanic and Atmospheric Administration (NOAA) satellites, tankers, and freighters can now plot their own position anywhere at sea on a digital onboard map. This might help avoid future Exxon *Valdez* disasters, and reduce accidents in foggy seas. The system may soon also help control congestion in commercial air lanes and locate stolen cars, any of which could be outfitted to beam a signal that would reveal their location.

Opposite top: Locust swarms like these pose another serious threat to crops in the Sahel.

Opposite bottom: Locust swarms can sweep across the country, devastating crops in their path.

The system can also be used to communicate as well as track. A trucker can be told via a hand-held liquid crystal display if there is a problem with the scheduling of his next shipment, and he in turn can let headquarters know if he's broken down in the middle of Kansas. Businessmen can now be contacted on the road no matter how far afield they go without having to constantly consult a telephone or check for messages or orders.

Some companies and municipalities are using commercial versions of the system to map areas by computer in a fraction of the time it would normally take. A transmitter is placed in a car or a truck, and as it travels the area it is mapping, it beams a continuous signal to a GPS satellite in geosynchronous orbit 22,300 miles above Earth. The satellite retransmits the signal to a computer that continuously plots the truck's location and draws a map as the vehicle travels. Any time the operator wants to note a landmark like, say, a fire hydrant, its location can be entered into a small computer and beamed up.

Future systems may work in tandem with experimental electronic maps in cars. A computer map displayed in the car can calculate the quickest route to a destination by checking with the satellite, which always knows where your car is in space, and then plot the best way to get there.

The satellite remains one of the great curve balls of the Space Age. Even pioneers like Arthur C. Clarke, John Pierce, and Verner Suomi admit they never foresaw the tremendous power and influence that these instruments would have. But in retrospect it isn't so difficult to see why they are so thoroughly transforming the world.

"Man is a communicating animal," says Arthur C. Clarke. "He demands news, information, entertainment, almost as much as food. In fact, as a functioning human being, he can survive much longer without food—even without water!—than without information."

When humans first emerged not so long ago, we were incapable of communicating with anyone out of direct earshot. Language was the first mechanism for sharing ideas, and the distance that our voices carried dictated the compass and extent of our

African farmers can use satellite maps to track locust swarms and save lives.

thoughts. Today satellites have made the reach of a single mind global and instantaneous.

Marshall McLuhan believed that all media, including the satellite, expanded and accelerated our biological capabilities. "[T]he wheel is an extension of the foot," he wrote, "the book an extension of the eye, clothing an extension of the skin, and electronic circuitry an extension of the central nervous system...When media act together they can so change our consciousness as to create whole new universes of psychic meaning."

In this sense the proliferation of satellites is as much a part of Earth's evolution as the making of mountains and the emergence of microbes. And the effects are no less profound. Earth is a planet being drawn together by the satellites that circle it. We are more in touch because of them, more aware, and more challenged.

Yet it is not altogether clear where satellite technology may lead. The ancients believed our futures are entwined with the stars, and it may be that this holds true somehow for these man-made points of light that we can see rolling across the evening sky each summer. In many ways the connections are no less mystical. There is, after all, more than a touch of magic in an instrument that flies and relays so many thoughts, ideas, and images with such speed and stealth, like telepathy.

The Agrhymet satellite station in the Niger Republic. The satellite dish looks out of place in a region largely untouched by modern technology, but helps to predict when famine due to drought or locust damage is likely. The station can predict bad conditions six months in advance, the time required by relief agencies to coordinate the shipment of food and aid. It is one of the more unexpected results of the Space Age.

6

"God, what a dancing spectre seems the moon."

—George Meredith

New Frontiers

Above: Visitors from another world. The moon may have had a hand in the evolution of the first life-forms on Earth. Some 3.8 billion years would pass before the progeny of those first cells dispatched samples of themselves to the lunar surface. Neil Armstrong took this picture of Buzz Aldrin during the *Apollo 11* mission.

Opposite: Astronaut Harrison Schmitt stands next to a huge boulder during the *Apollo 17* mission. Astronauts brought 845 pounds of lunar soil and rock back to Earth.

The Earth and the moon are two worlds cut from the same cosmic cloth. Current theory on the moon's origin holds that more than four billion years ago when the Earth was hot and battered, an object the size of Mars hurtled into it sending molten shards exploding into orbit. These fragments circled the planet until gradually, after countless collisons with one another, they combined to form Earth's one and only natural satellite.

When the young moon rose into the black and airless sky, it was colossal even in broad daylight. It hung a mere eleven thousand miles away, engulfing the horizon and circling the Earth every six and a half hours. Its great tidal forces pulled the planet's crust like taffy, and the two bodies swung together in space for eons before the moon eventually came to settle in an orbit 239,000 miles away. But its gravitational pull still beckons, and even today the motion of the tides generates enough heat to produce four billion horsepower.

Some scientists believe that the sloshing this tidal action caused in pools along Earth's ancient shorelines contributed to the evolution of the first cell. In the acidic seas of a primordial Earth, churned by the monstrous and circling moon, enveloped in a turbulent and noxious atmosphere, and cooked by the untempered radiation of the sun, life mysteriously arose. Or at least that is one theory.

Three and a half billion years later the progeny of these early cells dispatched samples of themselves across a quarter of a million miles of space in the forms of two Apollo astronauts named

Buzz Aldrin and Neil Armstrong. Clad in protective suits, they alighted on the surface of the moon and became the first Earthlings to walk upon the place that may have had a hand in their genesis. It represented one of the most difficult technological efforts humans had ever undertaken, but even without the machinations of Stalin and Khrushchev, Eisenhower and Kennedy, or the dreams of men like von Braun and Korolev, it would only have been a matter of time before we went there. The moon is simply too close, too ubiquitous, too thoroughly anchored in our psyche for us not to have journeyed there.

The earliest humans knew the moon as a child knows it, with glee and awe, troubled by its changeable shapes and reassured by its clockwork cycles. As long as we have kept records, we have praised, worshipped, and feared the moon; we have linked it with love, madness, and majesty. Primal societies considered the moon the true source of fertility: menstrual cycles parallel lunar cycles, therefore it seemed that the moon must somehow govern them. The same held true for the seasons: Planting and harvesting were set by the patterns of the moon. It was the giver of life.

For centuries, throughout the world, the chief deity for many cultures was the Moon Goddess, an all-embracing mother. Babylon's Ishtar is one of the oldest; the Phrygians worshipped Cybele, the Persians Anahita, the Celts Anu. Astarte, Aphrodite, Venus, even the Virgin Mary of Medieval Europe, all have their origins in these myths. They represent the mystery and power of genesis.

Ancient Greeks believed the waxing, crescent moon made crops grow and flocks increase as the moon itself grew to fullness, and its cycles came to mark the intervals between holy days (and still do in religions around the world). The Romans believed dead nobles walked its surface, and the Incas beat drums and sounded trumpets to reawaken it when it fell into eclipse. Long ago when we were separated and unaware of one another, cultures had the moon in common, and everywhere on the planet it became important to know when it would appear and what shape it would take.

Even today the moon remains at the center of our imaginations. We have conjured lunar creatures—bat-winged and insect-eyed. Full moons have made wolves of men and romantic lunatics of us all. We have crooned by its silvery light and traveled to its surface by the power of our imaginations, propelled by wind,

Above: An object the size of Mars collides with Earth. Current theory holds that this impact resulted in the formation of the moon as debris from the collision eventually coalesced in orbit around Earth.

Opposite: Plans for a super Saturn rocket that was to spearhead the settlement of the moon after Apollo. Had it ever been built it would have risen forty-five stories, one hundred and fifty feet higher than the *Saturn V,* which launched astronauts to the moon.

swans, dew drops, and a mystical substance called *lunarium*. As recently as the last century, newspaper articles reported telescopic observations of the daily lives of lunar creatures.

It made sense, then, that when we finally decided to depart Earth for another world, the moon was the destination, and in 1969, armed with a thirty-story rocket, several hundred tons of liquid hydrogen, kerosene, and oxygen, and the pooled talents of thousands, we *did* go.

The Apollo program was originally intended as a starting point for the human exploration of space, but instead it became a finishing line. In the early days of the space race, many assumed that the Apollo missions would be scouting forays, leading to an outpost, and then a colony. In 1966, three years before *Apollo 11* landed on the moon, NASA already had on its drawing board what it called the Extended Launch Vehicle—a super Saturn rocket forty-five stories high with extra engines strapped onto the first stage, which would be used to establish the first permanent lunar base after Apollo completed its initial missions. It was this kind of thinking that made a movie like Stanley Kubrick's *2001: A Space Odyssey* more plausible in the 1960s than it is today. But as Isaac Asimov later put it, "After *Apollo 17*, it became clear that the attitude toward expeditions on the Moon was 'We scored a touchdown. We won the game; now we can go home.'"

Before the first lunar missions, if you had asked anyone to predict the future, they probably would have suggested almost anything *but* the extraordinary events of Apollo—an organized fusion of human creativity, sweat, courage, and money; the development of astounding technologies; and finally a triumphal series of visits during which twelve humans left the prints of their boots in the lunar dust. "... [We] drank the wine at the pace they handed it to us," said one NASA veteran.

But after this first flurry of missions, nothing. No base, no observatory, not a single piece of hardware that pointed the way to the future. Selected areas of the moon lay strewn with various parts of six spaceships, six American flags, three rovers, several plaques, two golf balls, and one falcon feather. But twenty-four years after the last humans stepped on the moon it remains uncolonized, and untouched either by humans or by robots.

This is not to say that Apollo was nothing more than a dazzling stunt driven by the mania of the cold war. For the first time

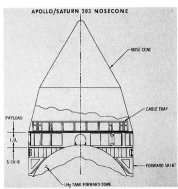

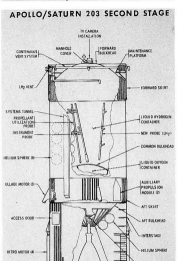

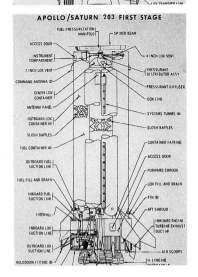

Above: A micrograph of a lunar sample. The properties of lunar rocks can be processed to produce building materials, oxygen for fuel and breathing, insulation, and even water.

Opposite: The Shimizu Space Hotel. The Shimizu Corporation hopes to build a resort hotel in space by the year 2020, a tall and expensive order. Hotel rooms circle two hundred feet from the main axis, creating an artificial gravity like that of Earth. Play rooms for weightless fun and games are located close to the axis where rotational force is not so great. Visitors could gaze down at the Earth or take day trips to the moon.

ever, human beings actually set foot on a heavenly body; the lunar missions revealed a new perception of our own planet that in turn revolutionized our attitudes about our role in Earth's future. Apollo moved the human race into the heady business of space exploration in the most dramatic way. And despite the political hoopla, important science was accomplished. Astronauts collected 845 pounds of lunar rock and dust and hauled it back to its place of origin to be analyzed for clues about the events that brought the Earth and moon into being, and during their brief forays, they planted the moonscape with geological devices that ranged from seismometers to heat-flow probes.

Yet the end of Apollo came almost as suddenly as its genesis. By 1970 the public had lost interest in the moon, and Congress had grown miserly toward NASA. Politicians complained that for the price of one moon shot the government could pay for countless new hospital beds or hire better teachers to educate the children of the United States. "If we can put a man on the .noon," people liked to say, "then why can't we _____," and then any number of phrases would fill in the blank: end crime, eradicate poverty, improve education, stop the war in Vietnam.

Of the six Apollo missions still under development in 1970, two were cut. On December 9, 1972, *Apollo 17* astronauts Eugene Cernan and Jack Schmitt landed in a lunar valley named Taurus Littrow, a place created when a meteor, or possibly two, slammed into the moon some 3.9 billion years ago. The astronauts remained five days, collecting samples and setting up instruments, and then on December 14 they headed back home, the last human beings to walk on moon dust and witness the Earth rising over the lunar horizon. No one seemed to care.

Despite this ebb in public interest, NASA remained hopeful that Congress, the American people, and the new Nixon administration would come to its senses and build upon Apollo's achievements. But no new lunar initiatives were funded. Our love affair with the moon seemed to have become merely a flirtation, a onenight stand inflamed by the passions of the cold war.

During the seventies and much of the eighties the moon continued to fall further out of favor. NASA occasionally triumphed with missions like Viking and Voyager, and in 1980 the shuttle finally flew, but none of these had any direct connection with the moon. Even the Soviet Union had dismissed the moon as a destination,

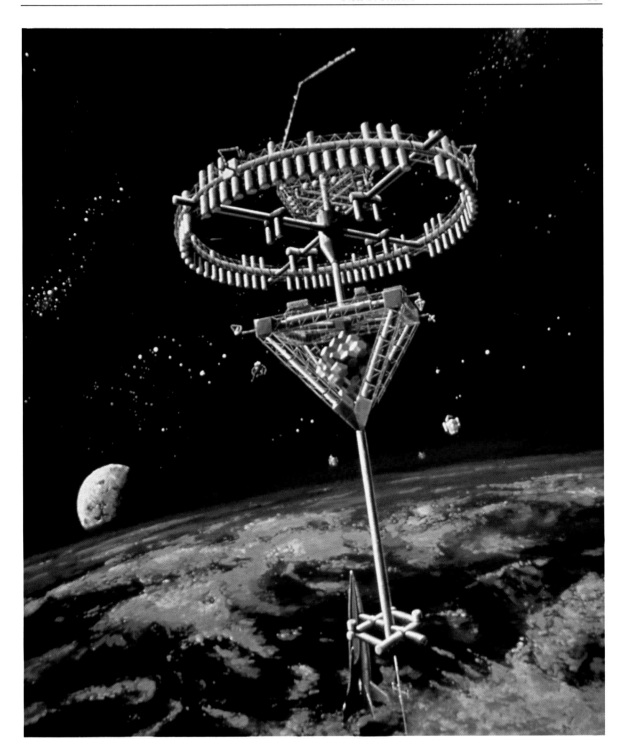

concentrating instead on their Salyut and Mir space station missions, apparently with a manned visit to Mars in mind. Sometimes magazines like the British Interplanetary Society's *Spaceflight* published an article that would briefly resurrect the possibility of lunar colonies, but the whole concept seemed to have developed the musty aura of a relic. The moon's future as a settlement had somehow become a fossil before it even existed.

But in the mid-1980s the National Commission on Space and the Ride Report gave the moon a new vitality. Both documents put pressure on the Reagan administration to revive the floundering space program. Each called for the establishment of manned bases on the moon and made it clear that any settlement should represent a long-term commitment, not a brief Apollo-style flirtation. The idea was that each step in the exploration of space should build upon the previous one, and the moon should be a centerpiece in a gradual expansion into space that would include an Earth-orbiting space station and human excursions to Mars. Once again, the moon had been raised to the status of a destination.

Today the drawing boards of the United States, Russia, and certain corporations in Japan are proliferating with moon bases. Japan recently dispatched a small probe that is now orbiting the moon and has scheduled a more ambitious mission that will jettison two miniprobes to the moon's surface in 1996. American and Russian scientists are developing spacecraft they hope will take a fresh look at the moon's surface and its geology in the 1990s.

What will become of the moon if these plans are pursued? Some scenarios involve commercial enterprises that will transform the lunar landscape into everything from extraterrestrial oxygen mines to luxury resorts. Others see the moon as a place for conducting scientific study—an unearthly Antarctica that belongs to no single nation, but serves as a free zone for science. Still others view the moon as a jumping-off point for the exploration of the solar system, particularly Mars and the various asteroids that wander near Earth. In this case the moon could also become a place for testing exactly how humans can successfully live in space. Being so close by, the argument goes, the moon is a perfect place to learn about the effects of low gravity on human beings, to test new technologies designed to protect against the dangers of radiation and isolation, and to try out fledgling efforts to create enclosed living systems.

Opposite top: In 1952 *Collier's* magazine published an article on the global advantages of the creation of a space station. The article raised America's awareness about the possibilities of space, but a station on this scale is still a futuristic concept.

Opposite bottom: A recent design for an orbital transfer vehicle (OTV) that would move heavy payloads from Earth orbit to the moon. Here the OTV uses its aerobrake to slow its descent into low Earth orbit before docking with the space station.

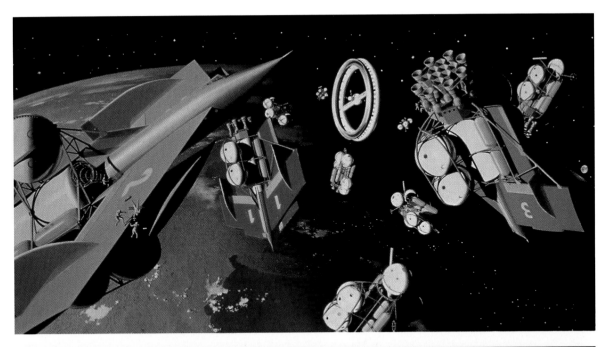

Wendell Mendell is among those who sees the futures of Earth and the moon as linked. He has the serene countenance of a bearded minister, but his belief is fierce. As chief scientist of Lunar Base Studies at NASA he sees the human race and Earth's only satellite as natural partners; for him, the moon stands at the headwaters of the solar system. Throughout the 1980s when others were yawning at lunar projects and talking about how to tap government funds for forays to Mars, he and a small corps of scientists and engineers were proposing the kind of lunar enterprises that later received so much attention in the reports of the National Commission on Space and the Ride panel. "We have got to look at space in a brand-new way," Mendell says. "Not so much in terms of projects and programs but as an evolving sector of our society and ultimately our economy." A lunar base seems sensible to Mendell because the moon is so nearby, a world for seriously testing the possibilities for the exploration of space.

In order to colonize the moon, rather than simply visit it, Mendell believes a reliable transportation system that can ferry people and supplies there is essential, an opinion von Braun himself would have appreciated. These lunar ships would require a space station that could operate as an Earth-orbiting dock, eliminating the need to launch expensive, multistage rockets like the *Saturn V* directly from Earth for each trip to the moon.

Some of the most memorable images in space pop culture are space stations: the dazzling wheeled contraptions designed by von Braun and popularized in films and television and by the artwork of Chesley Bonestell in the 1950s, for example. It was a descendant of these designs that rolled languorously around the Earth awaiting the arrival of the polite but secretive scientist Haywood Floyd in *2001: A Space Odyssey*. Though space stations of the future aren't likely to be as grandiose as those that von Braun envisioned, Mendell feels that they should serve as harbors in the sky, ports of transfer between ships arriving from Earth and others on their way to the moon. Ships from Earth, like the space shuttle or the so-called Heavy Lift Launch Vehicles now being considered to replace it, would make the steep journey from Earth to the station, and astronauts would then transfer to an Orbital Transfer Vehicle specifically designed to travel from the station to the moon. These would not be muscle-bound rockets like the old *Saturn Vs* or even the space shuttle; they would be smaller craft engineered for quick jaunts to the moon and back to the station.

Artist's conception of a space shuttle docking at the planned international space station. Four pressurized modules are located in the center of the structure. These modules will house a U.S. laboratory and habitation center, a European laboratory, and a Japanese laboratory. On the front surface of the truss: Canada's mobile servicing center and mobile manipulator arms.

This transportation system would not come cheaply, however. The gravity well out of which any rocket has to accelerate in order to leave Earth requires that it carry vast amounts of fuel at great expense—$3,000 per pound at present rates. For example, less than 2 percent of the shuttle is payload; the rest is the hardware and fuel required to enable it to leave Earth's gravity in the first place. When the shuttle blasts off, the gargantuan tank attached to its belly brims with hundreds of tons of the liquid oxygen needed to burn its hydrogen fuel. (Just as oxygen in the air makes it possible to burn a match, oxygen is required to ignite a rocket's hydrogen fuel.)

"If we launch everything from Earth in order to get to the moon and back safely, then for every ton we take to the moon, we would have to launch seven tons [of payload] from Earth!" says Mendell. In other words, provisioning the moon from Earth would mean that most of the mass making the trip would not be people or scientific instruments or cargo, it would be fuel. Luckily, says Mendell, a solution to this problem lies beneath the surface of the moon itself.

The moon is rich with rocket fuel; in fact, oxygen makes up 44 percent of its weight. Most of this oxygen lies in black and flattened areas known to early astronomers as maria, Latin for seas—ancient lava flows, like the one in Taurus Littrow created by enormous meteor impacts more than three billion years ago. *Apollo 11* landed in one of these basins, the Mare Tranquilitatis, the Sea of Tranquility, and discovered that it abounds with something called ilmenite, a mineral made of tangled molecules of iron, titanium, and oxygen. Scientists on Earth have already developed ways to extract oxygen from ilmenite, leaving behind the handy by-products of titanium and iron. If NASA acts on plans to establish a small base on the moon by the middle of the next decade, creating and deploying a machine that can make oxygen from loads of ilmenite would be among the first projects undertaken. If successful, it would mean that the moon would almost immediately begin subsidizing its own exploration. It will, however, take more than this one project to make the moon hospitable enough to warrant our return.

The moon has none of Earth's charms. It is gray and bereft of life. Its fourteen-day nights are unimaginably frigid, and its oven-hot

The Shuttle Remote Manipulator System, known as the "Canadarm" has been in use since 1981. It is 15.2 meters long, weighs 410 kilograms, and can move 30,000 kilograms in space either automatically or under astronaut control. It is used to place payloads into orbit, and to retrieve satellites for repair. Canadarm's most spectacular success was the rescue and repair of Solar Max, a $250 million solar observatory satellite.

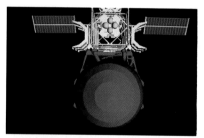

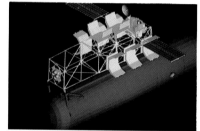

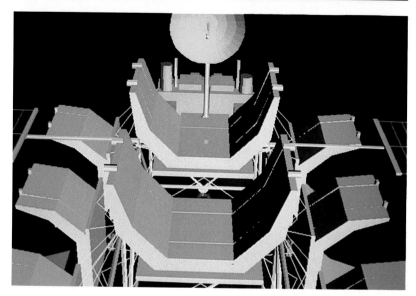

Global Outposts, in association with NASA, is developing a commercial space service platform. Launched by the space shuttle, the platform would provide power and communications in conjunction with the space station. The plan proposes that the external tank of the space shuttle could be recycled as a space platform simply by placing it in orbit instead of allowing it to fall from the shuttle toward Earth, where it breaks up upon reentry. In orbit the external tank could be used as the foundation for a privately financed space station that could be constructed using a remote manipulator system. The external tank weighs over sixty-six thousand pounds and is 154 feet long.

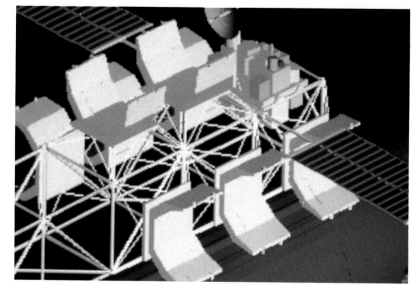

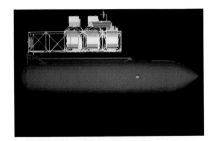
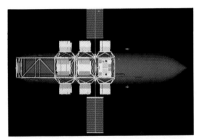
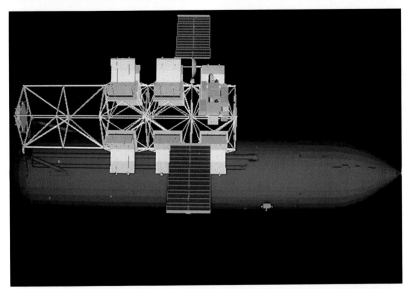
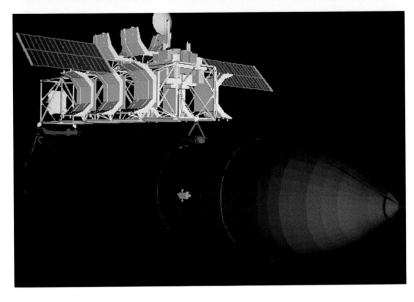

Above: The solar concentrator. The solar concentrator focuses solar power and can achieve temperatures of 2,300 degrees Fahrenheit. The concentrator can melt simulated lunar soil to create glass and other building materials. On the moon, the solar concentrator could be used to manufacture everything from building materials to household supplies.

Opposite: One artist's conception of an oxygen plant on the moon. Much of the moon's rock is bound up with oxygen, a key ingredient in rocket fuel.

Pages 270–71: The single eye of a future lunar telescope peers deep into the universe. With instruments this powerful there is no telling what secrets we might uncover.

days are just as long and unbearable. Nevertheless, Lawrence Haskins, a blunt-talking former chief of Planetary and Earth Sciences at the Johnson Space Center in Houston and now head of the Earth and Planetary Sciences Department at Washington University in St. Louis, looks upon the moon and sees abundance. To Haskins's mind, the moon suffers from a bad public image, but not one that a little creative science can't fix. During the fallow years following Apollo, the conventional wisdom was that we had been to the moon, found it airless, waterless, alien, and boring—a stingy, dead stone—and it made much more sense to get on with exploring the more inviting environment of Mars.

Thoughts like these "gore my ox," says Haskins. "We humans can do some pretty magnificent things with boring rocks and dust, like support human life and activities in space."

An average cubic meter of lunar soil, he points out, contains all of the chemical elements of a full lunch for two: two cheese sandwiches, two twelve-ounce sodas (with real sugar), and a couple of plums. "Of course there are a few steps of chemical separation and synthesis between a cubic meter of lunar soil and lunch, but I'm a rather lazy chemist and only interested in doing the separation of the elements. I'm willing to let organisms do the rest of the job, just as we do here on Earth."

Remaking the moon into useful products isn't an altogether new concept. As early as 1920 Robert Goddard suggested excavating the moon for fuel and other materials needed for space travel. And in 1950 writer Arthur C. Clarke suggested mining the moon, slinging chunks of it into lunar orbit from a machine that would use magnets to accelerate them along a rail, and then manufacturing the raw material into various futuristic craft and inventions.

These ideas happen to be very sensible. When Haskins and others scanned and cooked the lunar rocks returned to Earth by the Apollo astronauts, they found them loaded with all sorts of useful substances: oxygen, sand (or silicon), iron, calcium, aluminum, titanium, magnesium, and even traces of hydrogen. Given these raw materials, he says, humans can make a very nice living on the moon. Sand can be turned into solar cells to provide energy, and farther in the future silicon could be used to manufacture computer chips. (The frigid temperatures and airless lunar atmosphere are ideal for chip manufacturing.) Furthermore, the lunar rock can be used to cover and protect the first lunar habitats ferried from Earth. Three feet of it is enough to shield explorers

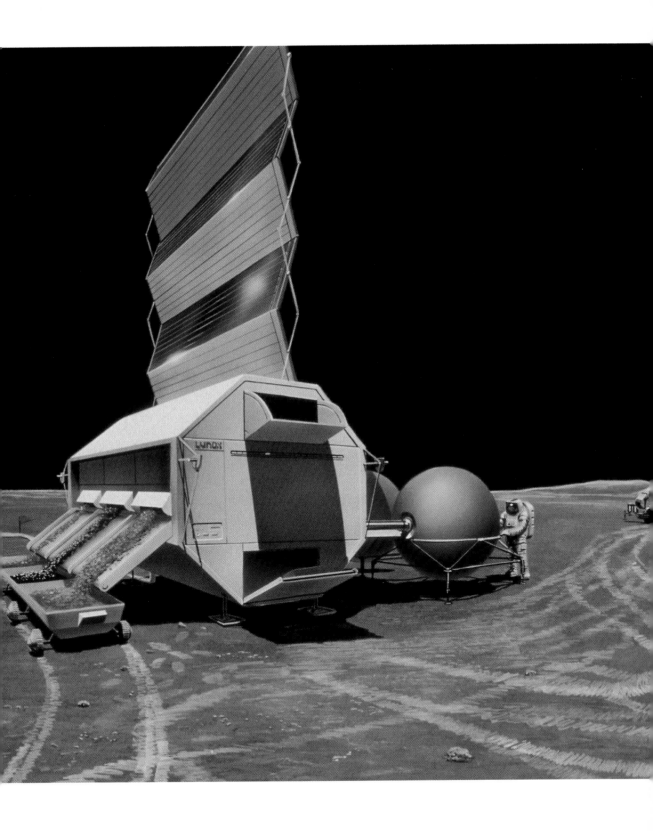

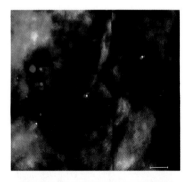

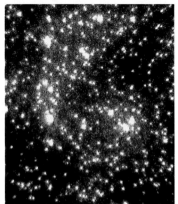

from radiation, micrometeorites, even the moon's intense heat and cold. The soil can be melted and cast into beams, rods, plates, tubes, and spun into superstrong glass fibers, which could withstand immense loads because the moon is so dry that there is no water to weaken their molecular structure. Yet even in this arid place, future astronauts could make water simply by adding hydrogen to oxygen. The chemical equivalent of a quart of water exists in every ton of lunar soil, and when combined with nitrogen, it can also be transformed into breathable air.

Haskins is willing to grant that only a geochemist might view the blasted body of the moon in quite this way; still, he wonders why we would want to travel 35 million miles to Mars when we can first test our mettle on a world only three days away? "I happen to believe," he says, "that sooner or later, in this country or in some others, old ideas and old knowledge are going to come together with new commitments, and the moon is going to become part of the sphere of human activity and influence."

NASA may be heading in this very direction. Following the recommendations made by the National Commission on Space, the Ride Committee, and the Bush administrations's Space Exploration Initiative, present proposals call not only for a permanent base on the moon but also a trial run scheduled for 2011 to test the feasibility of a manned mission to Mars. During that year, more than the image of a faraway Earth will be rising and falling in the black, lunar sky; a spaceship will also be orbiting overhead and will remain there for 200 days. The orbiting capsule will be a replica of the ship that would eventually take humans to Mars, and the six people inside would be undergoing an experience very similar to an actual Martian journey, spending nearly a year facing the dangers of space travel in the same cramped quarters. Following their mock journey, they will land on the moon, arriving soon after a cargo ship has supplied them with provisions and a temporary outpost, just as a similar ship would do on Mars. There they will remain for 40 days, roaming the surface in rovers, and finding out if they can still work with one another after such a long and difficult trip. They will also see if exercising on the bicycles and treadmills, or other contraptions provided to assuage the effects of weightlessness, has worked well enough to keep them up to the job of exploration. Then they will return to lunar orbit and simulate their long journey home, circling the Moon again, this time for 260 days more before finally returning to Earth.

One of the great advantages of carrying out this dry run on the moon is that during their mission the six would-be Martians will always have help nearby if necessary. (Plans call for a lunar base with a crew of twelve to have been established before the dry run.) In orbit or on the moon, an accident that cripples the food supply, for example, will not mean starvation, and if a hatch blows, a crew member is badly injured, or a key system breaks down, a rescue team could be immediately dispatched from the Earth or moon. If the trial run for Mars goes successfully, if the crew and cargo vehicles work properly, and the human beings themselves manage to survive one another's company, then NASA hopes to build on that mission and send six other astronauts on to Mars soon afterward.

Twenty-three hundred years ago Aristotle wrote that the cosmos was a place of perfection. The stars were baubles hung in a crystalline sky and the moon and sun were circular and smooth, not because of the cosmic forces that had formed them, but because the circle was considered the perfect form. It was a poetic view, but entirely incorrect.

Today astronomers no longer consider the moon a symbol of empyrean perfection but many feel that it could become the site of the perfect astronomical observatory. By a cosmic coincidence the moon rotates on its axis at almost precisely the same speed that it revolves around the Earth (about twenty-nine days). That means that only one side of the moon ever faces Earth, which leaves the other side perpetually hidden, but always face to face with the rest of the universe. This "dark side" (it is actually dark only half of the time) therefore makes a promising site for revolutionary discoveries.

We still picture the telescope as a great protuberance aimed into the night sky from the domed turret of an observatory, nothing more than an extension of the human eye, a gigantic high-tech version of Galileo's modified spyglass. But over the last forty years astronomers have found that there is far more to the universe than we can see with our eyes. The sky, it turns out, speaks in tongues, expressing itself across a wide spectrum of electromagnetic waves, and the visible part of the universe makes up only a small number of the stories the cosmos has to tell.

The Earth's atmosphere muffles a good many of the wave-

Opposite top: Image of material in the vicinity of a young star in the Orion Nebula, taken by the Hubble Space Telescope. Color is the result of different wavelengths of light being emitted by different elements. Light emitted by sulfur is seen as red, oxygen is shown as blue, and hydrogen as green. The red spots are globules of partially ionized hydrogen, and interstellar dust forms dark blobs. The sulfur emissions form structures the size of our solar system. Events like these may once have happened in our own solar system. If so, they eventually led to the human race.

Opposite center: 47 Tucanae, recorded by the Hubble Space Telescope's faint-object camera, is a globular cluster. When observed from Earth, only a few dozen of the stars it contains are visible, but the faint-object camera reveals several hundred. The brightest stars are called "blue stragglers" and are distinct from the red giant stars that predominate in this cluster. The blue stragglers may evolve backwards—changing from being cooler, older stars, back to hot, young stars. The high concentration of blue stragglers toward the center of the cluster suggests they are significantly larger than the other stars, and they could really be double star systems with the ability to influence the motions of thousands of the other stars in the cluster.

Opposite bottom: The Hubble Space Telescope at Lockheed Missile and Space Company. Light from stars and other celestial bodies is collected by the telescope's primary mirror. The light is then converted into digital radio signals that are relayed, through a Tracking and Data Relay Satellite, to a ground station at White Sands, New Mexico. Telescopes on the moon would greatly outperform even the Hubble Space Telescope.

Top: The Hubble Space Telescope being released from the payload bay of the space shuttle *Discovery,* April 25, 1990. The Hubble Telescope has created a Guidestar Catalogue, the largest inventory of stars ever recorded. It contains information on approximately 19 million objects. Optical telescopes on the moon could resolve images hundreds of times more clearly than Earth-orbiting observatories like Hubble.

Bottom: Hubble instrument replacement, scheduled for 1993. One of the instruments that may be replaced is the primary mirror. The mirror was ground to a curvature that is too shallow. As a result pictures are often fuzzy. Light rays that hit the outer edges of the mirror focus to a point that is about one inch away from the focus of light rays from the center of the mirror.

Opposite: Artist's conception of the Gamma Ray Observatory. The observatory weighs 15.3 tons and is the heaviest spacecraft deployed by the space shuttle. It is thirty feet long when deployed but expands to seventy.

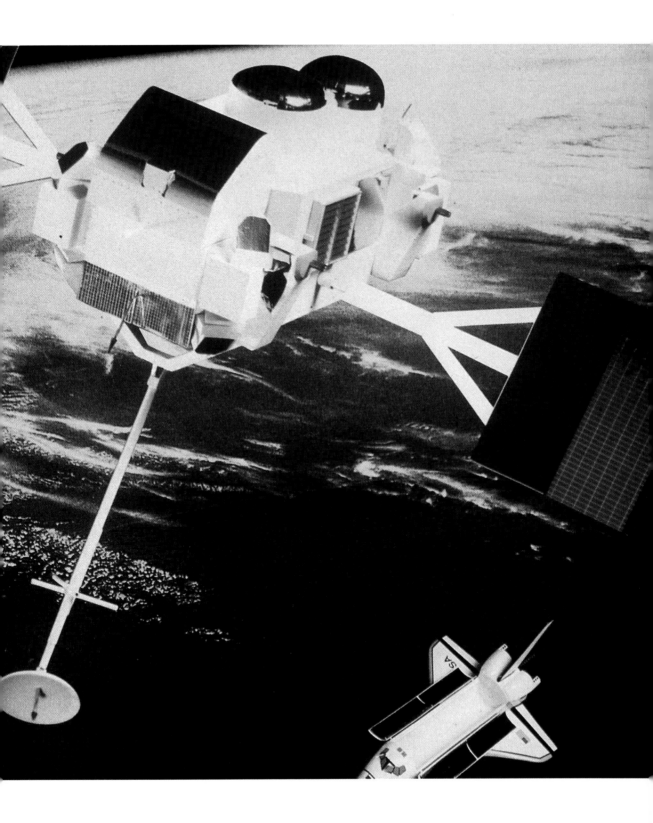

Above: The Very Large Array, in Socorro, New Mexico. The Very Large Array is composed of twenty-seven dish-shaped antennae that collect the low-frequency radio waves emitted by celestial objects. Each antenna focuses and collects the radio waves. These are then combined to simulate a single radio telescope view. Each antenna has an accuracy of plus or minus .05 of an inch and can focus on radio waves as short as .4 of an inch. The antennas can receive radio waves from any chosen point in the sky. None of them, however, can detect extremely low frequency radio waves. VLAs on the moon could.

Opposite: The Whirlpool Galaxy M51. This galaxy was first discovered in 1845 by William Parsons, Third Earl of Rosse. Spiral galaxies began their formation billions of years ago. After the Big Bang that theoretically created the universe, matter was distributed somewhat unevenly. The gravity of denser clumps, which became galaxies, attracted nearby gases. As these gases were pulled toward the galaxies, the galaxies began to spin, and flattened out along the axis of their rotation. Lunar observatories would reveal star clusters like these in far greater detail.

lengths in which the universe communicates, blocking X rays, gamma rays, even some radio waves. Astronomers have found that some of the most amazing phenomena in existence—quasars and plasma jets, pulsars and clues to black holes—are expressed in these hidden wavelengths. At one end of the spectrum are radio waves that are miles long; at the other end are gamma rays where oscillations rise and fall within the width of a single atom. Every one of these forces are part and parcel of the mechanics that drive the cosmic engine. Probes launched throughout the Space Age have already helped us to record some of these, but instruments on the moon would reveal them with unparalleled clarity.

The moon is free of an obscuring atmosphere; every molecule of its air could easily be pumped into a typical basketball auditorium. It is also quiet, utterly unperturbed by the radio noise that now blares from Earth, which was itself as quiet as a stone as recently as the 1920s. Furthermore, the lunar surface, unlike Earth's, is not given to rocking or trembling. Moonquakes, which are rare, shift the lunar surface no more than one billionth of a meter. This dearth of air, noise, motion, light, and heat all become advantages in the business of wresting secrets from the universe.

Moon observatories could improve upon almost every present-day method for peering into the sky, even NASA's Great Observatories, like the Hubble Telescope and the Gamma Ray Observatory that currently orbit Earth. These instruments vastly improve our views of the cosmos, but they are still plagued with problems created by Earth-circling debris, the planet's magnetic field, radio noise, and atmospheric drag. The largest radio telescope in the world is the Arecibo Observatory in Puerto Rico, which has a receiving dish that stretches a thousand feet across, and has been used to penetrate Venus's clouds and map some of its features by bouncing radio waves off its surface and reading the returning echoes. Yet, despite its great size, it "sees" radio waves no better than the human eye can resolve light waves. Enormous synthetic radio telescopes called VLAs, Very Large Arrays, have been built on Earth by setting up a series of antennas many miles apart with their dish-shaped sensors linked by a ganglia of cables, fiber optics, or satellites, which allow computers to create a single detailed synthetic picture. None of these telescopes, however, not even Arecibo or the largest VLAs, can detect extremely low-frequency radio waves because the planet's atmosphere blocks them. But in the silent, wide-open spaces of the moon,

Previous pages: End of a Two-Week-Long Lunar Day, a painting by Chesley Bonestell that first appeared in a 1952 edition of *Collier's* magazine. Although the moon is closer to Earth, in some ways it is less suitable for colonization than Mars. Mars has a cycle of light and dark almost identical to Earth's. On the moon each day and each night lasts for two weeks.

Top: Fornax A, Radio Galaxy. The central region of this galaxy exploded at some unknown time in the history of the universe, and two jets of matter were thrown out in opposite directions. The jets are composed of particles moving away from the center at about 60,000 kilometers per second, roughly one-fifth the speed of light. The particles are caught in a celestial backwater, creating lobes that continue to expand into the red clouds seen here. The energy in the center of this picture was created when this galaxy captured smaller neighboring galaxies.

Bottom: Cassiopeia A. Supernova Remnant. Cassiopeia A is the remnant of a star that may have died in 1680. At that time the outer layers of a massive star exploded and created a shell of the interstellar material around it. Expanding debris from deep inside the former star has finally caught up with this shell and is seen here breaking through it from inside.

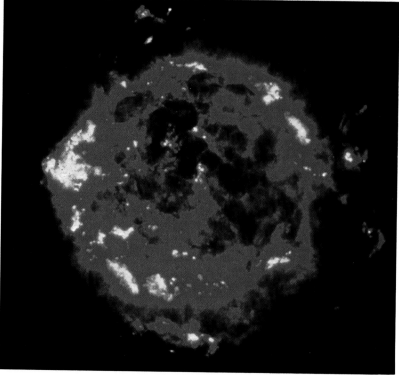

where no atmosphere exists, we might someday deploy lunar VLAs that can draw these radio waves down to the airless moonscape to reveal pictures of galactic clusters millions of light-years away, and open the door onto the origins of the universe.

Even the frigid lunar night is an asset. Everything in space gives off heat in infrared wavelengths, but it wasn't until the 1980s that astronomers began to investigate this aspect of the universe, partly because infrared telescopes have to be supercooled with liquid nitrogen so that they remain sensitive to the heat created by faraway objects. Using an uncooled infrared telescope on Earth would be like trying to make out the beam of a flashlight in a sunlit room. Craters at the poles of the moon, however, are always dark, which leaves them refrigerated to −250°F, perfect for an infrared telescope.

But understanding the potential of lunar observatories and actually building them are two different issues. Exactly how would we go about erecting telescopes on the moon? At first we would merely haul them from Earth and reassemble them on the lunar surface. Eventually, however, to reduce costs, we would have to find ways to fabricate instruments right out of the raw material of the moon using the methods favored by Lawrence Haskins. We could design small factories to smelt moon rock into aluminum and titanium, from which we could fashion the skeletons of observatories. The moon's bone-dry silicon could also be turned into excellent telescopic mirrors, as well as spun into the structures that support them. Under these circumstances the moon's low gravity is an advantage because even in the case of a gargantuan observatory, these structures would only have to support a fraction of the weight that they would on Earth.

With instruments this powerful, anything might come into view, perhaps even another planetary system. Though there have been tantalizing clues, right now scientists can't say with certainty that other solar systems even exist; no telescope, not even Hubble, has glimpsed any planets that may have formed around another sun. But an array of optical lunar telescopes could resolve faraway objects ten thousand times better than Hubble, perhaps a star of medium size like our own sun, with a planet like Earth circling it on which life has somehow emerged. Somewhere else events may have already occurred that will lead to telescopes one day being trained on us, like those of H. G. Wells's Martians.

Still another possibility is a lunar radio telescope that will

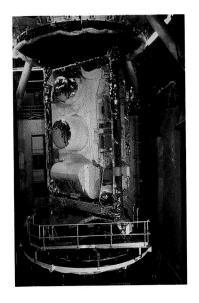

The Gamma Ray Observatory in its test chamber at TRW Inc. These instruments are the most advanced and sensitive ever flown. They study gamma ray sources from ten to fifty times fainter than any previously sensed. This expands our ability to observe the universe by 300 percent. Gamma ray radiation is a form of energy that is invisible to the naked eye, but is present in the universe and observable by these delicate sensors. Gamma rays are emitted during the most basic of the processes of the universe, like the births and deaths of stars, and the formations of black holes, quasars, and pulsars.

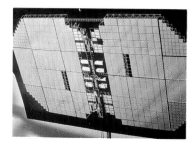

allow us to succeed where code expert William Friedman and others failed, plucking out of the cacophony of the universe extraterrestrial noises that make sense, signals that perhaps indicate a knowledge of prime numbers, or sounds as beautiful and ordered as the music of Brahms or Mozart; messages from far away that would say, in effect, "We are here. You are not alone." The interstellar conversations that follow would be tentative, with long pauses, but they would mark the beginning of a new age, and remind us again that we are not, as we sometimes think, the center of the universe.

Whichever plans for building outposts on the moon are carried out, early efforts will likely have the feel of a company town or the drilling settlements of Alaska. Although Earth will support the first forays to the moon, settlers will eventually have to fend for themselves and pay their own way with whatever they can extract from the lunar soil. The creation of oxygen to supply fuel for ships traveling between Earth and lunar space will likely anchor this early economy. These fuel supplies could be launched into space and transported in tugs to an Earth station where workers would construct and maintain ships as they do in any port. Later much of the raw material for building the ships themselves could be mined on the moon and transported to Earth-orbiting space harbors. The income from this trade could help subsidize the expansion of the moon's observatories, its life sciences experiments, and even fledgling business ventures, perhaps in the field of microgravity technology. A large network of solar arrays might someday be built on the moon, creating a high-tech version of photosynthesis with energy from the sun warming, cooling, and powering large closed ecological systems supplied with plants and animals from Earth, and nurturing the first lunar farms and ranches. These new ecosystems would enclose all lunar life within a single living loop, not only for growing food, but for creating and recycling air and water. Slowly life will begin transforming the moon the way seeds make a forest or corals create a reef. It may happen, however, that the moon will never reach this stage of development. This kind of growth, after all, requires very strong financial incentives. The moon may simply become an extraterrestrial scientific reserve, a vast observatory, or a test bed for forays to Mars—a world of trial runs and technological assessments. If the moon is to become the next locus of human migration, however, it will need to develop a broad-based economy.

A space solar power station in Earth orbit. Similar stations could collect energy from the sun and microwave it back to Earth for conversion to conventional power sources.

An artist's conception of a solar power satellite station. This satellite would provide five thousand megawatts of electrical power to a ground receiving station. The solar collection area is fifty square kilometers, and the total weight of the satellite is fifty thousand tons. Our planet only receives one billionth of the energy of the sun. They would be very expensive to build, but solar satellites could harness some of this energy, providing an alternative to the use of fossil fuels.

The foundation of the first colonists' economy in the New World five hundred years ago was the gold and silver stolen from the civilizations of South and Central America. It was an ignominious beginning, but later agricultural trade developed, and as settlers streamed across the Atlantic to the Americas, an even richer and more varied trade economy with Europe evolved (often heavily subsidized by corporate/government cartels like the Hudson's Bay Company or the Dutch West India Company). That process created an economic symbiosis that transformed both worlds, old and new, in profound and unpredictable ways.

In return for staples and manufactured supplies, the New World colonists shipped furs, lumber, and fish across the sea, and soon Europe came to rely on products from America. Crops grown by American natives—corn, peppers, potatoes, tomatoes, chocolate, coffee—revolutionized diets around the world. Economic power in Europe shifted according to a nation's ability to control shipping and to finance colonies. The experience of new lands and cultures blossomed into revolutionary political ideas and philosophies. Rousseau's doctrine of the noble savage arose from the discovery of native Americans, and very possibly the heads of French monarchs would never have rolled in Paris if thirteen American colonies hadn't declared independence from Britain thirteen years earlier. Democracy was little known in Europe or anywhere else five hundred years ago; now it is one of the world's dominant forms of government. Strange and unpredictable events follow in the wake of discovery.

Could similar economic and social marriages develop between Earth and the moon? Looking upon the moon as we have, through the eyes of astronauts, it is hard to understand how a place so dead might become an integral partner in Earth's economy, but it might.

Someday, for example, the moon could export star power. During the lunar day the sun's energy flows to the surface of the moon unimpeded by atmosphere and uninterrupted for fourteen consecutive Earth days. The solar arrays that power small outposts in the future could be expanded across large sections of the lunar landscape and the energy exported to Earth. Tsiolkovsky, the first to write of harnessing the power of the sun for use on Earth, would have appreciated this idea. As he pointed out, our planet receives only one billionth of the sun's energy; the rest simply dissipates into the unending blackness of the universe.

Top: A mass driver on the lunar surface. Construction of solar-powered satellites is an expensive venture. The mass driver could make it possible to mine the moon of raw material that would then be launched into space and remanufactured into solar satellites.

Bottom: The mass driver can be used as a reaction engine as well. Here, a mass driver propels an asteroid through space by using bits of the asteroid as fuel. The reaction from the material tossed into space propels the asteroid to its destination. The parts not used are transported to a space station to be processed into building materials, water, and gas.

Solar power beamed from space was seriously considered in the mid-seventies following the Arab oil embargo. NASA and the United States Congress investigated building solar-powered satellites (SPSs or "sunsats") in orbit around the Earth, but the cost was considered too great. The National Academy of Sciences calculated that constructing a flock of sixty satellites over fifty years' time to power the United States would run up a bill of three trillion dollars.

Princeton physicist and author Gerard O'Neill, however, has argued that the moon could greatly reduce these costs by providing the metal and glass needed to build the satellites. The raw materials, as Arthur C. Clarke had suggested in the 1940s, would be launched from the lunar surface with a magnetic rail gun, or as O'Neill calls it, a mass driver. Such a machine could theoretically sling small chunks of the moon, mined by robots, in a continuous stream into lunar orbit where they would be collected by a simple orbiting factory that would then process the small payloads into their basic elements, and transport the resulting iron and nickel and titanium to full-scale space factories in orbit around Earth. (O'Neill, also the founder of the Space Studies Institute in Princeton, has even built a working mass driver that can accelerate a small bucket of material from 0 to 300 miles per hour in less than a second.)

O'Neill has argued that this approach would dramatically cut the cost of assembling parts of the satellites on Earth and then hauling them up into orbit, because slinging the raw material from the moon takes so much less energy and fuel. Once built, the satellites' solar cells would gather energy from the sun, convert it to microwaves, and beam it to receiving stations on Earth.

Critics point out that building SPSs in space with raw material from the moon would require the development of highly complex and untested technologies, and that transmitting microwave energy to Earth could be dangerous to the wildlife that falls within the path of the beam. Nevertheless, in recent years both Soviet scientists and Japanese firms like Shimizu and Mitsubishi have seriously investigated sunsat technology. So far, however, nothing has advanced beyond the drawing board.

Some scientists believe that a rare isotope called helium 3, which exists in abundance on the lunar surface, may make the moon the source of a different, more revolutionary, kind of energy. Helium 3 came to its lunar resting place because of the

Two worlds that may once have been one, the Earth and the moon, could again be drawn together in the future if we chose to colonize the moon. Here an Orbital Transfer Vehicle (OTV) from a space station near Earth rides into orbit around the moon. Such vehicles could become the trade ships of the future, not unlike the trade ships that followed Columbus to the New World.

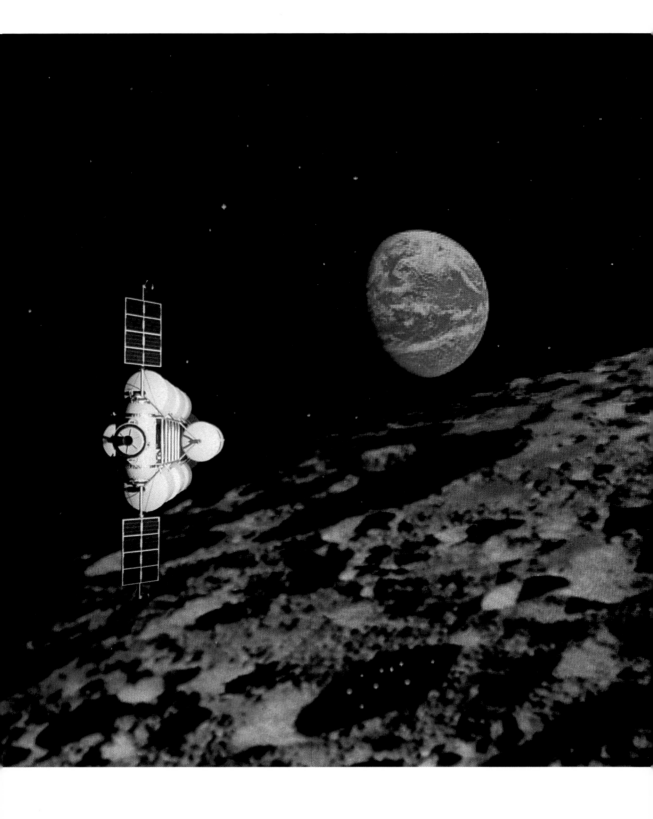

ongoing nuclear reaction within the sun, which generates billions of tons of helium and sprays it into space on the solar wind; for 4 billion years this stardust has been falling on the moon, while Earth's thick atmosphere has blocked it from settling on its surface. The reason helium 3 is so intriguing is because it may be useful as a fuel in nuclear fusion reactors, which mimic the same process that powers all stars.

Scientists have understood the process of nuclear fusion for more than fifty years but still haven't succeeded in implementing it. The problem is that a fusion reactor must be capable of generating core temperatures of 200 million degrees F, yet be small, efficient, and controllable. Otherwise, it would annihilate everything on the planet. In effect it has to act as a *tiny* sun, a contradiction in terms.[10] The standard fuel used in fusion reactor tests is tritium combined with a form of hydrogen extracted from water called deuterium. But helium 3 from the moon would have advantages over tritium as a fusion fuel. It is three times less radioactive and theoretically easier to fuse, and although it is more difficult to ignite, some feel it could cut the time needed to develop workable fusion reactors by as much as a decade.

The development of a practical fusion reactor would represent the greatest leap in energy technology since the evolution of photosynthesis, and could arguably make helium 3 the most valuable export in the solar system. There is enough helium 3 on the moon to provide for the energy needs of the Earth and the moon for thousands of years.

Whether or not moon-made solar power satellites or helium 3 actually become real lunar exports in the future, they illustrate how a lunar economy that links the futures of the moon and Earth might evolve and make large-scale lunar settlement a sensible endeavor, just as the products of North and South America accelerated their colonization and joined the societies of the Old and New Worlds.

But no matter how valuable lunar products become, large-scale lunar migrations will never begin until people have a safe, affordable way of getting there. So far the cost of traveling in space is outrageously expensive, and remains the exclusive province of astronauts handpicked by government agencies. With no more

Large unpiloted cargo ships like these, each based on the design of the space shuttle, could carry payloads much heavier than those presently transported by the shuttle.

Bottom: The *Ariane 4* rocket rises from its launch pad in the jungles of French Guiana. Arianespace was the first company to enter the commercial launch services market. The *Ariane 4* is a modular launch vehicle designed to launch payloads of up to ninety-four hundred pounds. The vehicle can place from one to three satellites in a highly precise orbit. It made its maiden voyage on June 14, 1988.

---

[10]   In November 1991, scientists in Oxfordshire, England, produced a significant amount of power from an experimental fusion reactor called JET. It produced 1.5 to 2 million watts of electricity in a two-second pulse, a major breakthrough. Before this the most that had been created was a few thousand watts.

than a handful of shuttle flights per year, each carrying an average of five passengers, and a price tag of $2 billion per shuttle, it costs $400 million to fly one shuttle-launched human into space, per flight—not the sort of price that sends people rushing to the ticket counter.[11] Millions travel all around the world in jets not because the technology has become any simpler, but because they can afford it. Even after paying $150 million for a jet, it costs an airline only an average of $100 per passenger per flight to move them from one place to another. If it hadn't been for large ocean-going sailing ships that made passage to America affordable, the colonists who followed Columbus might easily have opted to remain in the Old World until something better came along. The same holds true for the rough efficiencies of the Conestoga Wagon or the railroad. Without these, who can say how long it would have taken to win the American West?

Today the only ships that can launch people into orbit are NASA's shuttle and the Soviet Union's Proton and Energia rockets. The European Space Agency (ESA) is developing Hermes, a mini-shuttle, and Japan is creating HOPE, a similar, unpiloted vessel. All of these essentially use the same basic rocket technology that von Braun's engineers developed at Peenemünde half a century ago. They remain expensive and problematical, high-tech hot rods, nothing like a Boeing 747.

In the past five years, however, there has been a renewed push to investigate ships that can take off horizontally, like a jet, and regularly fly passengers directly into space. This concept is similar to ones investigated by Russian rocket pioneer Fridrikh Tsander and considered by Stalin in 1948 when he was searching for an Earth-orbiting bomber. Stalin turned the idea aside because he thought it was too difficult, and today daunting technical hurdles still remain.

Germany and the United States are farthest along in researching the potential of space planes. (Japan also has a prototype on its drawing boards, but hasn't publicized many details of its work.) Both U.S. and German designs have their roots in efforts undertaken before the launch of *Sputnik*. Germany calls their space plane Sänger in honor of Eugen Sänger, the inventor who pioneered the ship's piggyback design. Sänger, an Austrian aeronautical genius, was a contemporary of von Braun's, and it was his

---

[11]   If the development costs associated with the shuttle are not included, the cost drops to $60 million per passenger.

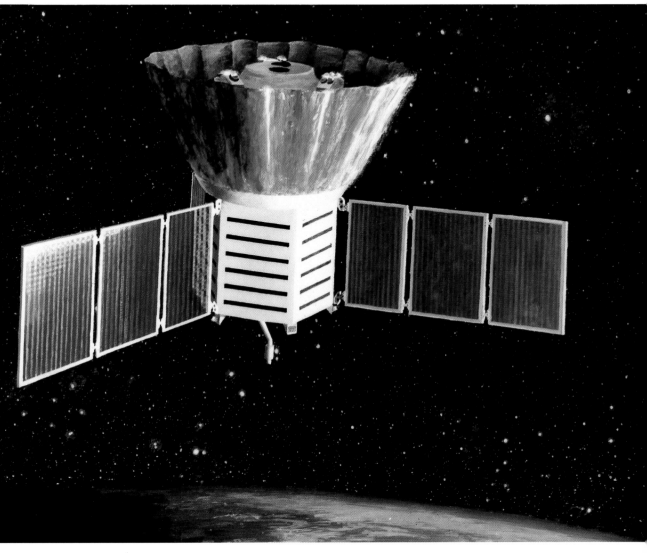

COBE, the Cosmic Background Explorer, is a space probe designed to measure background radiation in the sky and help shed light on the origin of the universe. In the spring of 1992 it revealed evidence that the universe was indeed created some fifteen billion years ago following a "big bang" which resulted in the creation of time and space from an infinitely small, hot and dense point of matter that eventually evolved into the universe of galaxies and stars that exists today.

brainchild that Stalin briefly considered for his bomber.

EMPIRE OF THE RISING MOON: JAPAN'S SPACE PROGRAM

After the United States, Russia, and Europe, Japan has the largest and most successful space program in the world. Unlike the United States, which runs its civilian program under NASA's enormous technological and budgetary umbrella, Japan's is much smaller and divided into two separate agencies: NASADA (National Space Development Agency) and ISAS (Institute for Space and Astronautical Science). NASDA is commercially oriented and has been overseeing the launch of communications, broadcast, and remote sensing satellites as well as the development of Japan's new launch rocket, the H-II. It is also developing HOPE, a small automatically piloted craft similar to the shuttle that will be able to ferry crews and supplies to space station Freedom and then return to Earth.

ISAS is a small and highly academic organization that has launched simple but very efficient and effective scientific missions to Halley's comet and the moon, and conducted highly successful astrophysics experiments. Its reputation for solid and creative science and engineering allow it to take a meager budget farther than most space agencies.

In addition to these two organizations there are Japan's powerful and visionary conglomerates, corporations such as Mitsubishi, Shimizu, Obayashi, and others, which are considering enormous space projects: resorts in Earth orbit and on the moon, lunar scientific bases, orbiting solar-power satellites, and enormously complex colonies on Mars. So far these projects haven't advanced any farther than the drawing board, but the thinking is ambitious, impressive, and innovative.

A Sänger space plane would be two ships really: one, a supersonic plane powered by a revolutionary engine called a ramjet, the other a smaller rocket that sits on the back of the ramjet. The ramjet would take off from a designated runway somewhere in Europe, then swoop down over Africa to the equator where it would make a sharp turn east to gain a kick from the rotation of Earth. When it reached twenty-one thousand feet and a speed of Mach 6 (six times the speed of sound or 4,500 miles an hour), it would jettison the smaller, passenger-carrying rocket from its back. This rocket would then fire its engines, accelerating to Mach 25, or 17,700 miles per hour, the speed necessary to escape Earth's gravity. When the Sänger is ready to return to Earth, it could simply coast in like the space shuttle and land on a runway.

The United States' effort is called the National Aerospace Plane (NASP), otherwise known as the Orient Express or the X-30. It is the most ambitious aeronautical project in the world, and if successful, will be able to take off from an airport and rocket at Mach 25 directly into orbit. The X-30's pedigree goes back to the X-15 project of the early sixties, the Right Stuff, fighter-jock days at Edwards Air Force Base in Southern California when pilots like Chuck Yeager and Joe Walker were taking jets up over the salt flats of the Mojave desert and flying them higher and faster than any other vehicle of their day. But in the heat of the space race, the X-15 project was eventually mothballed in favor of the manned Mercury missions, which would put a man into space sooner. X-15 pilots had flown close to fifty miles high, nearly a quarter of the way into low Earth orbit, but they never came anywhere near Mach 25—the jet's engines simply couldn't attain that speed.

The Aerospace Plane proposes to solve this problem by incorporating three engines into one craft. All jet engines require oxygen to burn their fuel, but at very high altitudes the air grows so thin that they suffocate. When the space plane would take off, a conventional turbojet like those that power a standard supersonic fighter would accelerate it to Mach 3. Then a ramjet would kick in. A normal jet engine uses its turbines to force air through it rapidly enough to burn the fuel at a rate that allows the plane to maintain its speed. But a ramjet doesn't use a turbine to compress the air because the 2,200-mile-an-hour speed of the jet is already ramming enough oxygen into the engine—at least until the jet reaches Mach 6.

This next stage requires an exotic piece of technology called a scramjet, essentially a more efficient, cooler-running version of the ramjet, an engine that can force the ever-thinning molecules of air through itself fast enough to enable the plane to reach Mach 25. Only then will the Aerospace Plane have generated enough speed to fling itself into orbit.

A full-fledged space plane could be the vehicle that spear-

heads the peopling of space because theoretically it would be much less expensive to operate than anything ever built. It is completely reusable (not partly reusable like the shuttle), doesn't require expensive launch facilities, and most of all, it doesn't have to haul its own oxygen. When each shuttle takes off, it tugs 650 tons of liquid oxygen along with it; the X-30 would leave that weight on the ground and draw its oxygen from the air.

But it will be some time before a space plane flies; although its proponents hope to have tested a prototype before the end of the decade, the technical hurdles are so great that such a flight might not be likely to occur until some time in the next century.

Assuming that the pooled imaginations of humans can hurdle the problems of reaching space cheaply and safely, and assuming we are able to settle the moon, what would it be like?

Not many people have considered this question as thoughtfully as physicist and engineer Krafft Ehricke. Ehricke was one of the more gifted and visionary engineers who worked on von Braun's rocket team during World War II. As early as 1942, while at Peenemünde, he investigated the possibilities of nuclear rockets, and before that he had won patents on two rocket inventions. After the war he played a leading role in the development of the Atlas rocket in the United States and is considered the father of the Centaur rocket. As early as 1965 he began considering what he called the "extraterrestrial imperative"—his belief that the human race would eventually explore and then evolve beyond the world of its origin.

Ehricke died in 1984, but moon specialist Wendell Mendell once recalled that after he first heard him speak on the future of the moon, he felt like "a student who had been working long and hard on a homework problem, and then had suddenly been shown the answer in the back of the book."

Ehricke believed that it was not only logical but inevitable that the human race would colonize the moon. But he also agreed that migrations there would never take place until people had a compelling reason to set down roots. We would have to create opportunity on the moon, he believed, and develop a lunar economy that could provide a variety of goods for Earth.

In the 1970s he felt that Earth was running out of energy, and that the human race was approaching the practical limits of planetary growth. To him this was similar to the dilemma life on Earth

Top: The model X-15 experimental rocket-powered aircraft. The X-15 has been flown to an altitude of 354,200 feet at 6.7 times the speed of sound.

Bottom: The X-15 in flight.

had faced two billion years earlier, when after more than a billion and a half years of growth, single-cell organisms in the early seas began running out of food. As the situation reached a crisis, evolution produced the chlorophyll molecule, which enabled new organisms to harness energy from the sun and use photosynthesis to create a new source of food. From that moment on, life exploded across the planet, leading eventually to complex organisms, human beings among them.

For Ehricke, photosynthesis represented an astounding technological breakthrough. Earth was no longer, strictly speaking, an altogether enclosed system. These new forms of life had expanded the biosphere to include the sun, and by tapping its power became, as Ehricke put it, "astrogenic." By reaching beyond the planet, life had turned an absolute limit on growth into a limitless ability to grow, an advance that resulted in the creation of vast amounts of oxygen (a by-product of photosynthesis) and led to the creation of the biosphere that Earth has today.

Ehricke believed that colonizing the moon represents a new way for life on Earth to extend its astrogenic reach. By stretching beyond our planet, he hoped we might turn a limit on growth into an evolutionary opportunity, just as the single-cell microbes had.

Like many others he saw energy as the first lunar export, but also foresaw a fascinating panoply of other lunar products, envisioning four potential markets for a complex lunar economy. The first market would be the inhabitants and institutions on the surface of the moon itself. Lunar mining operations would supply these pioneers with raw material for lunar observatories, spare parts, and habitats. Lunar utilities would turn lunar rock into oxygen and hydrogen, supply fuel, water, and nuclear and solar energy. Other businesses would engineer and manufacture precision instruments and materials for living and working on the moon.

The second market for moon products would be the space around Earth itself. By the year 2000 Ehricke estimated that two thousand satellites would be circling the planet in geosynchronous orbit, 22,300 miles high, instruments that would by then be indispensable to the social and economic life of the planet. This orbital world would require precision instruments and raw materials for satellite maintenance and construction, services that could be more easily and less expensively provided from the moon than from Earth. The moon could also provide a cheap supply of fuel, ships, and people to do their work, and when it came time for a satellite to be replaced, lunar businesses could easily retrieve and salvage or recycle them.

Ehricke also foresaw a similar market in low Earth orbit, about two hundred miles above the surface of the planet. Already the United States, Japan, and Europe have discussed the construction of microgravity laboratories and light manufacturing facilities to be established there, and the Mir space station actually operates a lab that conducts these kinds of experiments. The European Space Agency plans to launch an orbiting Man Tended Free Flier and an American company, Space Industries International, has similar plans to launch its Industrial Space Facility—both are small laboratories that astronauts will visit periodically while they work and live on the proposed Freedom space station. Ehricke felt that someday in the future all of the needs of these low-orbiting stations could be more quickly and cheaply supplied from the moon than from Earth.

The fourth and largest market for lunar products would be Earth itself, a market for which Ehricke imagined fascinating exports. The moon, for example, might be able to supply certain products made from rare alloys and powders that lunar factories could manufacture more efficiently because of the moon's low gravity or frigid temperatures, or because it would be so much easier to reach zero gravity manufacturing sites in space from the moon than from Earth. The moon might also become a supplier of unique information and experiences for humans on Earth. An educational and entertainment industry might grow up around the moon's astronomical observatories. Images of the cosmos never before seen might be piped into homes all over the Earth via satellite, and people could journey electronically to the sites where the first astronauts had set down fifty or sixty years earlier, without ever leaving their living rooms. A homegrown special effects industry could blossom. Film and television producers wouldn't have to build sets or rig cables to create a sense of weightlessness or imitate the terrain of an alien world, capabilities Fritz Lang or Steven Spielberg would no doubt appreciate.

The moon's low gravity and exotic locale would also entice investors to create a resort and tourism industry in which people could play low-gravity golf, or bound fifty feet in the air on trampolines, or float over the surface of the moon in low-energy scooters. Eventually this industry would lead to a vast culture as

independent from Earth in the future as the Americas are from Europe today.

The inhabitants of Ehricke's culture would live in lunar cities that he named androcells—transparent enclosures miles long and sixteen hundred feet high in which mirrors would bend sunlight to provide illumination during the lunar night, and special filters would help imitate the hues of Earthly sunshine. He called this settlement Selenopolis after the Selenite creatures that inhabited the moon in H. G. Wells's science fiction adventure *The Men in the Moon*. Though they or their forebears may have come from Earth, the inhabitants of Ehricke's androcells would just as certainly be creatures of the moon. They would live and work there, and would build biospheres that would grow eventually to house hundreds of millions. Environments of all sorts would honeycomb it, moonlike, Earthlike, and surrealistic; places only an imagination could conjure.

The Selenians would become, as he put it, the "Cosmopolynesians" of the solar system, futuristic Ionians, an itinerant and curious people without ties, who effortlessly travel the islands of the inner solar system, transforming their varied experience into new visions and new knowledge.

It is a long way from the present to Selenopolis, but the plans we now have in mind for modest lunar bases and a halting investigation of the moon's barren surface may be a sign that the primal, beckoning force of the moon is inexorably drawing us back to it to study the possibilities imagined by Ehricke, Tsiolkovsky, and others. Princeton physicist Freeman Dyson, inspired by the science fiction of writer Olaf Stapledon, has imagined highly advanced civilizations that have harnessed the power of whole stars and encased their entire solar system in a blazing nest of technology. In contemplating a return to the moon we may be starting out in this very direction, reclaiming what was once a part of our planet. Perhaps by the turn of the next millennium, the Earth's rocket-born attenuations will mark the genesis of one world reaching out ever so carefully to another in cosmic embrace.

Shimizu Corporation's plan for a lunar base constructed of cement hexagons. The hexagons can be connected at any point to form fully pressurized structures.

Following pages: The sun sets on Mars, August 20, 1976, as the *Viking I* lander looked out upon the western horizon of Chryse Plantatia, the Plains of Gold. Will we someday populate many worlds just as Tsiolkovsky predicted?

EPILOGUE

"The essential wildness of science as a manifestation of human behavior is not generally perceived."

—Lewis Thomas

The Shape of Things to Come

Above: Two Earthlings step foot on Mars in the twenty-first century.

A cargo ship orbits Mars awaiting the arrival of humans on their way from Earth, more than 35 million miles away. Once they reach Mars, the cargo ship will be the key to the crew's survival.

For aeons life was Earthbound. Now, in the form of the human race, it has found a way to leave the planet and expand its frontiers into space. Never before have we had more than a world to explore: Today we have an entire cosmos. Not entirely by design, the knowledge and experience gathered during the Space Age has made us more capable of resolving the fundamental questions that initiated it in the first place: Where do we fit in? How did we come to be? What forces power the universe? Are we alone?

Searching for the answers to questions this fundamental is bound to lead to unknown territory: contact with other life-forms, for example. In the seventeenth century Christiaan Huygens, a physicist of ferocious intelligence and impeccable credentials, and his more famous contemporary, Isaac Newton, both speculated on the likelihood of life on other worlds. Huygens wrote in his book on cosmology, entitled *Cosmotheoros,* "A man that is of Copernicus's opinion [that the sun—not Earth—is the center of the solar system], . . . cannot but sometimes think that it's not improbable that the rest of the Planets have their Dress and Furniture, and perhaps their Inhabitants too as well as this Earth of ours." Newton speculated that just as the microscope had found all sorts of strange life in everything from water to blood, "so may the heavens above be replenished with beings whose nature we do not understand." Even Tsiolkovsky wrote a book of science fiction entitled *Dreams of Earth and Sky* that speculated on alien life.

Being a social species, we seem reluctant to be alone. The idea of extraterrestrial life was once considered material for pulp science fiction and B-movies, but today scientists take the possibility very seriously. As large as the universe is, with an estimated 100 billion galaxies, and a daunting 300 billion stars in the Milky Way, our chances of having company look promising. In a physics experiment as vast as the universe, the processes that brought life into existence are not likely to be unique to our small world. We may find that life is as common in cosmic evolution as brains and eyes are in the evolution of life on Earth. Scientists do not yet have any hard evidence, but there are indications that many of the stars we see each evening in the sky are surrounded by planets just as our own sun is, some of them places where life might have arisen only recently, others where life-forms may have been evolving billions of years longer than humans and could already be busily colonizing their galaxy.

Many scientists believe that the best chance we have for discovering these other civilizations is to listen for them using radio telescopes around the world. It is possible that just as our television signals are strong enough to drift out into space, an alien civilization is leaking bits of *its* culture into the cosmos too. Or perhaps they are as curious to know if they are alone as we are, actively sweeping their sky with a radio beacon the way a New England lighthouse sweeps the sea. If so, we have our ears cupped and our radio satellite dishes tuned in, systematically searching the universe for the call of other beings far away.

But this is not as simple as it sounds: the universe resounds with radio noise. Everything from stars to hydrogen molecules emits a message in thousands of frequencies that together make a sound like wind. On rare occasions astronomers have found signals that disturb these interstellar breezes, occurrences so strange that those listening have allowed themselves to think, for a single astonishing moment, that they had actually stumbled across another version of intelligent life. But when they re-aimed their radio telescopes at the origin of the signal, it always disappeared, nothing more, perhaps, than a stellar or molecular aberration in the normal whoosh of the universe.

Princeton physicist Freeman Dyson believes we can vastly reduce the territory that radio telescopes scan for radio signals by searching not for a message but for heat. In his book *Disturbing the Universe* he writes that finding evidence of a civilization at our own level of advancement, one struggling toward successfully managing the resources of its planet, would be difficult; there would be few clues apparent across intergalactic distances. We might, however, have a chance of finding evidence of what he calls a type 2 civilization, creatures who have harnessed the power of their sun. "There is one kind of emission a type 2 civilization cannot avoid making. According to the second law of thermodynamics, a civilization that exploits the total energy output of a star must radiate away a large portion of this energy in the form of waste heat . . . as infrared radiation, which astronomers on Earth can detect."

Instead of these approaches, however, some have suggested we should dispatch probes in search of extraterrestrial life. Robert L. Forward, a respected engineer at Hughes Research Laboratories, who, like Tsiolkovsky before him, puzzles on the problems of spaceflight and dabbles in science fiction, believes that ships designed to cover interstellar distances reasonably quickly could carry robotic probes to other star systems.

Interstellar travel, he admits, is a monumental undertaking. "It is difficult to go to the stars. They are far away and the speed of light limits us to a slow crawl along the starlanes." Although Hollywood has conveniently outmaneuvered the speed of light, in the real world where the laws of physics apply the cosmic speed limit is set at 186,300 miles per second, and despite the determined efforts of many scientists, no immediate methods for skirting the regulations have been found.

186,300 miles per second seems an incredibly fast speed until you consider the distances involved. The moon, for example, is a quarter of a million miles away and takes three days to reach in a chemical rocket. Neptune, the eighth planet from the sun, is ten thousand times farther away than the moon, and with great difficulty, and after more than a decade, Voyager managed to arrive in 1989. But these distances are piddling compared with the territory that lies between Earth and even the nearest star, Proxima Centauri, which is 4.3 light-years away, ten thousand times farther than Neptune and one hundred million times farther than the moon! *Pioneer 10,* the swiftest object human beings have ever built, has now cleared our solar system and is hurtling into interstellar space at 25 miles a second—an impressive 90,000 miles an hour. But that is still 7,500 times slower than the speed of light. It would take *Pioneer* 33,000 years to reach Proxima Centauri, and 15 billion years, the present age of the universe, to reach the nearest galaxy! We are a long way off from even a "slow crawl."

But assuming that we could invent ships able to attain even one-third the speed of light, we would then be capable of reaching seventeen star systems inside of forty years, well within the span of a single human lifetime. According to Forward the technologies that could make this possible are not very far off—nuclear pulse engines driven by hydrogen bombs, fusion rockets, even ships fueled by antimatter are possible, if dauntingly expensive.

As unreal as its name sounds, antimatter, when combined with an equal amount of matter, releases 200 percent of its mass as energy. Scientists have already manufactured antimatter, but so far only in vanishingly small amounts, a few subatomic particles at a time. Antiprotons are produced and stored at the European Centre for Nuclear Research (CERN) in Switzerland. Difficult to make, and even more difficult to store, antimatter has to be slowed, cooled, and trapped by lasers or magnetic fields that suspend it, like a trussed-up prisoner. Furthermore, the estimated cost of producing it in the twenty-first century is $10 million a milligram—$9.1 quadrillion a ton.

On the other hand, it wouldn't take very much antimatter to power a spaceship. Once developed, such vessels would be able to sail from one end of the solar system to the other in days or weeks, rather than years, making a journey to Mars hardly longer than a cruise to Europe. Sending one-way probes to the nearest star systems in this manner would reduce travel times from millennia to a mere twelve or thirteen years.

However, as with all fuels, antimatter has its shortcomings. The laws of relativity hold that as an object approaches the speed of light, its mass likewise approaches infinity. Truly high speeds, therefore, require enormous amounts of fuel to move the ever-increasing weight. If an antimatter-driven space shuttle could be built that traveled, say, 99 percent of the speed of light, it would require a half million tons of antimatter. The solution to this problem is to invent a ship that doesn't need to carry along its own fuel, one that grazes the galaxy for atoms as it travels. In 1960 physicist and engineering consultant R. W. Bussard designed just such a vessel. He called it an interstellar ramjet, a ship that looks something like a plunger with an enormous saucer at its nose, which could scoop up hydrogen atoms floating in space and accelerate them into a fusion reactor that would then propel the ship at speeds approaching light. This scoop would have to be hundreds of miles across because in interstellar space, matter is rare—about one atom every ten cubic centimeters. Grazing the wide open spaces between stars would require a very large mouth.

To say that Bussard designed the ramjet isn't quite accurate because that would imply it could be built, but it can't. Theoretically, it would work, but the technology needed to assemble such a leviathan and sophisticated machine is beyond anything science can tackle right now or is likely to be able to tackle for many years. The design is really a thought experiment rather than a blueprint, and it raises some interesting issues about high-speed travel, because if Bussard's ramjet could be built, it would amount to a true time machine. A crew who boarded it and traveled from Earth toward the center of the Milky Way at close to the speed of light would arrive twenty-one years later according to their on-board clock, but the laws of relativity stipulate that thirty thousand years will have passed on Earth. Following a return trip, a total of sixty thousand Earth years would have passed, yet the crew would debark from their ramjet only forty-two years older. The planet they return to would be an unimaginably different place than they departed, and the human race, assuming it had survived, thoroughly changed. (To illustrate how much can change in sixty thousand years, anthropologists still debate whether or not true human beings even existed on Earth that far in the past.)

If we do make contact, who or what might we find? Physical appearances are a wide-open question. In *Dreams of Earth and Sky,* Konstantin Tsiolkovsky imagined creatures with glassy, airtight bodies and winglike appendages that enabled them to live in a vacuum and use sunlight to manufacture food. (It's interesting to note that Tsiolkovsky, who was deaf, had the creatures communicate in this vacuum visually, not sonically, with brightly colored images that constantly changed beneath their transparent skin.) Robert Forward has written a science fiction book in which jellyfishlike aliens live on a water planet. Some scientists have imagined creatures that have no solid bodies, but exist as pure thought, as patterns of magnetism.

Looks aside, are aliens likely to be sinister like H. G. Wells's Martians or wise and peaceful like Percival Lowell's? Some scientists believe any truly advanced civilization must be wise and benign, otherwise they would very likely have wiped themselves out, just as we might yet do. Long progress would tend to indicate that they have overcome not only the desire to annihilate themselves but also the desire to spread annihilation to other quadrants of the universe. War is costly; it might not be possible to simultaneously slaughter one another *and* develop the advanced technologies that enable travel to other solar systems. In our case, the cold war may have catapulted us into the Space Age, but it also damaged it, channeling vast amounts of money and human talent in the United States and the old Soviet Union into multibillion-dollar weapons systems. On the surface it may seem that evolution has tended to favor predators, creatures that are fast, flexible, and aggressive, but the ultimate inclination of living things seems to be to cooperate and build. Life on Earth has made the planet a monumental and cooperative construction project carried out by millions of species—symbiosis on a planetary scale. Seen this way, predation might only be a smaller part of the general process of cooperation; rampant and perfectly efficient predators, would ultimately leave any planet with only one surviving species and nothing else to eat, except for one another.

Assuming cooperation of an earthly sort is common throughout the universe, any aliens we contact may well be willing to talk. But establishing lines of communication with extraterrestrials isn't likely to be easy no matter how wise and happy they are to share their knowledge and experience. How, after all, can we expect to communicate with creatures whose reality might be as different from ours as ours is from a termite? We already know that dolphins and certain whales have large, complex brains, but we haven't yet found a way to communicate with them except on the most basic level. Could we find a common language with another race from another part of the universe?

Astronomer Carl Sagan believes that mathematics is our best shot at a cross-cosmos lingua franca; a knowledge of prime numbers, for example, could help establish a system of communication. In his book *The Lives of a Cell,* Lewis Thomas suggests a little more whimsically, that the best way to explain ourselves to extraterrestrials might be in the music of Bach. "We would be bragging of course," he writes, "but it is surely excusable for us to put the best possible face on at the beginning of such an acquaintance."

Even if we found a way to communicate, what would we talk about? If the creatures we contact are highly advanced, maybe we could ask whether they have solved any of the many difficult questions that continue to elude us: Is there a way to exceed the speed of light? Have you seen God? Is there sadness on your world? Are children a universal joy? Astronomer Frank Drake, now a professor at the University of California at Santa Cruz and a man who pioneered the radio-telescope approach to extraterrestrial searches, feels that even a rudimentary exchange could be of immeasurable benefit. A simple yes or no to the question "Will fusion power work?" would save the human race billions of dollars and years of effort.

Beyond all of these concerns, simply making contact would absolutely rattle the foundations of every culture on Earth. Wholesale reassessments of science, religion, and philosophy would be in order, a situation that might leave us temporarily speechless. But then, perhaps a little silence would be best.

"After all," Lewis Thomas writes, "the main question will be the opener: Hello? Are you there? If the reply should turn out to be Yes, hello, we might want to stop there and think about that, for quite a long time."

Of course there is the possibility that we *are* alone in the universe, in which case the Space Age, rather than producing a means for making contact, is more likely to focus on spreading human life around as Tsiolkovsky imagined nearly a hundred years ago. We could set out for other galaxies in states of suspended animation, reduced to the "granular state" that Goddard imagined in his view of the far future, "The Ultimate Migration." Or, to defeat the obstacle of long intergalactic journeys, we might digitally download our minds into a futuristic computer, as Hans Moravec, a computer scientist at Carnegie Mellon University in Pittsburgh, has suggested, a machine that will not run down like a biological body and brain and could make it possible to beam ourselves in the form of a laser from one galaxy to another.

We are already beginning rudimentary work of this kind as we study the possibilities of permanently inhabiting the moon and Mars. For now we are content to build shelters or outposts on these worlds, burying ourselves beneath their surfaces like seeds, but not very far in the future we might begin to terraform these places, rendering them Earthlike.

Mars is the most probable candidate for this kind of planetary metamorphosis since at one time it may have been warm, volcanic, and watery like Earth. Today it is altogether different—frozen and nearly airless, barren and pocked, but the old, basic constituents for life may well be locked in Mars's rusty rock. Looking at it this way, says Mars expert Chris McKay at NASA's Ames Research Center, "[T]erraforming could be viewed as a restoration project."

Terraforming suits Mars in other ways: Mars tilts at almost precisely the same angle on its axis as Earth does, has frozen water and plenty of carbon and oxygen, and the length of both planets' days are nearly identical. The ultimate terraformation of Mars, however, would require the same ingredient that so successfully wrought the terraformation of our own planet. "The only known mechanism that can change a planetary atmosphere is life," says McKay. "At a very early stage we might release on Mars specially engineered microorganisms that could adapt to the extreme cold. As the planet warms and the atmosphere thickens, hardy species of grasses and shrubs might be introduced, followed by flowering plants, trees, and food crops." This aspect of the plan takes a page out of Lynn Margulis's theories. Given a basic environment and life's habit of filling whatever niches it can, microbes could take over the job of transforming Mars just as they did on Earth more than three and a half billion years ago.

Prompting the propagation of Earthlike planets in this way will also change the human beings who come to live upon these new worlds. Like the varieties of finches Darwin noticed on the Galápagos Islands, over time the settlement of other planets and moons will breed variations on us. Within a few generations, Mars would beget Martians, not humans, creatures branching from an Earthly evolutionary line, as surely as human ancestors branched from the same line that gave rise to the chimpanzee and great ape. By extending this process to other worlds, the solar system might someday become a cosmic rain forest, brimming with new species of plants and animals and intelligent beings, all springing from common terrestrial ancestors.

Not all of the central quests of the next age look to the future of life; some look back. The nature of our origins lies hidden among the wild percolations that produced the universe in the first place, phenomena completely unknown before we developed the ability to leave Earth. As scientists begin to understand celestial manifestations like pulsars, quasars, and the formation of galaxies, they are inching closer to uncovering nature's ultimate secrets: how space, time, matter, and life came about.

The Space Age led to the discovery of the X-ray universe and many of the other universes that lie along the electromagnetic spectrum. Orbiting observatories like the Cosmic Background Explorer (COBE) and the Hubble Telescope are now joining the minds of scientists with these instruments in space, further pulling back the curtain on the universe and the forces within it. Orbiting observatories belie the perfection and absolute celestial order that Aristotle said he saw when he looked into the sky. Modern astronomers and astrophysicists know today that the universe is anything but serene. Beyond the Milky Way, for example, a swarm of thousands of galaxies amounting to trillions of trillions of stars are hurtling across the universe at 700 miles per second, beckoned by some siren center of gravity. Astronomer Alan Dressler of the Carnegie Institution of Washington calls it the Great Attractor. "A structure so gargantuan deserved an equally good name," he says. The best guess is that the Great Attractor is an immense collection of galaxies 100 million light-years thick and 200 million light-years wide. Its mass may equal the mass of 10,000 trillion suns.

Some surmise that the universe is perforated with holes that are black and invisible, so dense that the unimaginable power of their gravity swallows everything around them—light, matter, all evidence of being. Where they exist, the pressures and forces at work may be so crushing that the laws of the universe—what we normally call reality—cease to apply. The "here" of a black hole is nowhere, and literally anything is possible. Elsewhere astronomers have found galaxies colliding, stars igniting like fireworks, and quasars where the bodies of entire suns explode as they are shredded and sucked wholesale into black holes. In other places there are walls of galaxies woven into lattices hundreds of millions of light-years long, and out in the endless night may be unknown dimensions; cosmic strings that have warped the structure of the cosmos from the beginning of time; even wormholes, passages to empyrean gardens where whole parallel universes may be in bloom. We may live, in other words, not simply within one universe but among an infinitude of infinite universes.

The forces that drive these phenomena are the same ones that power the cosmos, and created the sun and the Earth, which have in turn brought forth the human race. The ingredients of life, it seems, are contained in the stars; we are their by-products, percolations of the universe cooked up into consciousness and let loose to comprehend it. But even though the Space Age represents a new level in our long effort, our comprehension is still only in its earliest stages.

In the end, is there any sense to what we are up to? Where, after all, is the logic in building rockets and space stations, searching for life on other worlds, and launching observatories that cost billions to gaze deep and wide-eyed into the matrix of space and time? Why do such outlandish undertakings so thoroughly engage our wonder? Apparently we have little choice. We are congenitally curious, a trait that has proven invaluable over the long haul, keeping us constantly prospecting for new knowledge. Without this wanderlust, without our ability to dream and wonder, we might still be squatting in damp caves without even a fire to warm us or a rock in hand out of which to strike a tool. As it is, our curiosity has now catapulted us beyond our own world.

Nevertheless, having found frontiers beyond Earth (having proven the sky is *not* the limit), it remains difficult to say precisely what will come next. The more questions we answer, the more it seems mystery befalls us. The Space Age has upped the ante of exploration. It represents a branch point in our evolution, like the first lungfish that crawled up out of the sea onto land. We may not be able to explain precisely why we are departing the familiar surroundings of one environment for the dangers of another, but now we have taken the step, and such adventures inevitably foreshadow momentous change. The results are likely to be surprising and far beyond anything we can imagine right now.

After all, of the creatures that first slithered tentatively from the sea, which of them could have foreseen a dinosaur, an ape, or a human, must less a computer, a rocket, or a robot?

Bibliography

General

Bronowski, Jacob. *The Ascent of Man.* Boston: Little, Brown, 1974.

Burke, James. *Connections.* London: Macmillan, 1978.

————. *The Day the Universe Changed.* Boston: Little, Brown, 1985.

Campbell, Joseph, with Bill Moyers. *The Power of Myth.* New York: Doubleday, 1988.

Dyson, Freeman J. *Disturbing the Universe.* New York: Harper & Row, 1979.

Ferris, Timothy. *Coming of Age in the Milky Way.* New York: Morrow, 1988.

Harvey, Brian. *Race into Space: The Soviet Space Programme.* New York: Halsted Press, 1988.

Kennedy, Paul M. *The Rise and Fall of the Great Powers: Economic Change and Military Conflict from 1500 to 2000.* New York: Random House, 1987.

Margulis, Lynn, and Dorion Sagan. *Microcosmos.* New York: Simon and Schuster, 1986.

Oberg, James E., and Oberg, Alcestis R. *Pioneering Space: Living on the Next Frontier.* Foreword by Isaac Asimov. New York: McGraw-Hill, 1986.

Sagan, Carl. *Cosmos.* New York: Random House, 1980.

Thomas, Lewis. *The Lives of a Cell: Notes of a Biology Watcher.* New York: The Viking Press, 1974.

United States. National Commission on Space. *Pioneering the Space Frontier: The Report of the National Commission on Space.* New York: Bantam Books, 1986.

Wilford, John Noble. *Mars Beckons: The Mysteries, the Challenges, the Expectations of Our Next Great Adventure in Space.* New York, Alfred A. Knopf, 1990.

Chapter 1. Dreams to Reality

Boorstin, Daniel J. *The Discoverers.* New York: Random House, 1983.

Bronowski, Jacob. *Magic, Science, and Civilization.* New York: Columbia University Press, 1978.

Costello, Peter. *Jules Verne: Inventor of Science Fiction.* London: Hodder and Stoughton, 1978.

Epstein, Beryl Williams, and Samuel Epstein. *The Rocket Pioneers.* London: Lutterworth Press, 1957.

Goddard, Robert Hutchings. *The Papers of Robert H. Goddard, Including the Reports to the Smithsonian Institution and the Daniel and Florence Guggenheim Foundation.* Esther C. Goddard, editor. New York: McGraw-Hill, 1970.

Jensen, Paul M. *The Cinema of Fritz Lang.* New York: A. S. Barnes, 1969.

Kosmodem'ianskii, Arkadii Aleksandrovich. *Konstantin Tsiolkovsky.* Moscow: General Editorial Board for Foreign Publications, Nauka Publishers, 1985.

———. *Konstantin Tsiolkovsky: His Life and Work.* Translated from the Russian by X. Danko. Moscow: Foreign Languages Pub. House, 1956.

Lehman, Milton. *This High Man: The Life of Robert H. Goddard.* Preface by Charles A. Lindbergh. New York: Farrar, Straus, Giroux, 1963.

Ley, Willy. *Rockets, Missiles, and Men in Space.* Newly revised and expanded edition. New York: The Viking Press, 1968.

Ott, Frederick W. *The Films of Fritz Lang.* Secaucus, N.J.: Citadel Press, 1979.

Riabchikov, Evgenii Ivanovich. *Russians in Space.* Edited by Nikolai P. Kamanin. Translated by Guy Daniels. Prepared by the Novosti Press Agency Pub. House, Moscow. Garden City, N.Y.: Doubleday, 1971.

Tsiolkovsky, Konstantin. *Beyond the Planet Earth.* Translated by Kenneth Syers. New York: Pergamon Press, 1960.

———. *Collected Works of K. E. Tsiolkovsky.* Edited by A. A. Blagonravov. Washington, D.C.: National Aeronautics and Space Administration. For sale by the Clearinghouse for Federal Scientific and Technical Information. Series title: NASA technical translation F-236-238.

———. *Selected Works.* Compiled by V. N. Sokolsky. Edited by: A. A. Blagonravov. Translated by G. Yankovsky. Moscow: Mir Publishers, 1968.

———. *Works on Rocket Technology.* Edited by M. K. Tikhonravov. Washington, D.C.: National Aeronautics and Space Administration. For sale by the Clearinghouse for Federal Scientific and Technical Information, 1965. Series title: NASA technical translation F-243.

Winter, Frank H. *Prelude to the Space Age: The Rocket Societies, 1924–1940.* Washington, D.C.: Smithsonian Institution Press, 1983. Published for National Air and Space Museum.

Chapter 2. The Explorers

Bergaust, Erik. *Wernher von Braun: The Authoritative and Definitive Biographical Profile of the Father of Modern Space Flight.* Washington, D.C.: National Space Institute, 1976.

Clarke, Arthur C., ed. *Project Solar Sail.* Introduction by Isaac Asimov. New York: Roc, an imprint of Penguin, 1990.

Diamond, Edwin. *The Rise and Fall of the Space Age.* Garden City, N.Y.: Doubleday, 1964.

Dornberger, Walter. *V-2.* Translated by James Cleugh and Geoffrey Halliday. Introduction by Willy Ley. New York: The Viking Press, 1954.

Emme, Eugene Morlock. *A History of Space Flight.* New York: Holt, Rinehart and Winston, 1966.

Fajardo, Mario E. "Limitations on Stored Energy Densities in Systems of Separated Ionic Species." Paper accepted by AIAA, *Journal of Propulsion and Power,* May/June 1991.

Friedman, Herbert. *The Astronomer's Universe: Stars, Galaxies, and Cosmos.* New York: W. W. Norton & Co., 1990.

Harvey, Brian. *Race into Space: The Soviet Space Programme.* New York: Halsted Press, 1988.

Koppes, Clayton R. *JPL and the American Space Program: A History of the Jet Propulsion Laboratory.* New Haven: Yale University Press, 1982.

Lamont, Lansing. *Day of Trinity.* New edition. New York: Atheneum, 1985.

Lebedev, Valentin Vital'evich. *Diary of a Cosmonaut: 211 Days in Space.* Translated by Luba Diangar; edited by Daniel Puckett and C. W. Harrison; foreword by Michael Cassutt. College Station, Tex.: PhytoResource Research, Inc., Information Service, 1988.

Ley, Willy. *Rockets, Missiles, and Space Travel.* New York: Viking Press, 1951.

McDougall, Walter A. *The Heavens and the Earth: A Political History of the Space Age.* New York: Basic Books, 1985.

Oberg, James E. *The New Race for Space: The U.S. and Russia Leap to the Challenge for Unlimited Rewards.* Harrisburg, Pa: Stackpole Books, 1984.

Oberth, Hermann. *Die Rakete zu den Planetenräumen.* Munich: R. Oldenbourg, 1923.

Ordway, Frederick Ira, III, and Mitchell R. Sharpe. *The Rocket Team.* New York: Crowell, 1979.

Pohl, Frederik, and Frederik Pohl IV. *Science Fiction Studies in Film: Science Fiction Goes to the Movies.* New York: Ace Books, 1981.

Shelton, William Roy. *Soviet Space Exploration.* New York, Washington Square Press, 1968.

"Solar Sailing and the World Space Foundation." Paper privately published by World Space Foundation. South Pasadena, Calif.: World Space Foundation, 1991.

Von Braun, Wernher, and Frederick Ira Ordway III. *History of Rocketry and Space Travel.* New York: Crowell, 1967.

————. *Space Travel: A History.* Fourth edition, revised in collaboration with David Dooling. New York: Harper & Row, 1985.

Walters, Helen B. *Hermann Oberth: Father of Space Travel.* Introduction by Hermann Oberth. New York: Macmillan, 1962.

Winter, Frank H. *Prelude to the Space Age: The Rocket Societies, 1924–1940.* Washington, D.C.: Smithsonian Institution Press, 1983. Published for National Air and Space Museum.

Chapter 3. Quest for Planet Mars

Allen, Joseph P., with Russell Martin. *Entering Space: An Astronaut's Odyssey.* New York: Stewart, Tabori & Chang, 1984.

Biro, Ronald R., ed. *The 1991 NASA Space Life Sciences Training Program Curriculum Workbook.* Prepared for the NASA/KSC Payload Projects Management Office and Biomedical Operations and Research Office, Kennedy Space Center Florida by The Bionetics Corporation.

————. *The 1989 NASA Space Life Sciences Training Program: Part 4, Controlled Ecological Life Support System Program.*

Bluth, B. J. "The Psychology and Safety of Weightlessness." Paper presented at the 15th Symposium on Space Rescue and Safety, International Astronautical Federation, Session II, Space Safety and Rescue, Paris 1981.

———. "Soviet Space Stress." *Science 81* (September 1981): 30–35.

———. "Space Station in the 21st Century: A Social Perspective." Paper from the conference "AIAA Space Station in the Twenty-first Century," September 3–5, 1986, Reno, Nevada.

Bluth, B. J., and Martha Helppie. *Soviet Space Stations as Analogs.* Second edition, with Mir Update: May 18, 1987. Report of NASA Grant NAGW-659. August 1986.

The Boeing Aerospace Company. "Space Station Habitability Report." Submitted by National Behavior Systems, February 28, 1983.

Brewer, George R. *Ion Propulsion: Technology and Applications.* New York: Gordon and Breach, Science Publishers, 1970.

Broad, William J. "Space Station Mir Is Nearing Completion." *The New York Times,* March 13, 1990, eastern ed., B5.

Burgess, Eric. *To the Red Planet.* New York: Columbia University Press, 1978.

Burrows, William E. *Exploring Space: Voyages in the Solar System and Beyond.* New York: Random House, 1990.

Carr, Michael H. "The Steps to Mars." Presentation at the AASS Annual Meeting, Rationale for Human Exploration of Mars, February 18, 1991.

Casani, John, et al. *Proceedings of the AIAA/JPL 2nd International Conference on Solar System Exploration.* Pasadena, Calif.: California Institute of Technology. August 22–24, 1989.

Collins, Michael. "Mission to Mars." *National Geographic,* November 1988, pp. 733–764.

Committee on Human Exploration of Space, National Research Council. *Human Exploration of Space: A Review of NASA's 90-Day Study and Alternatives.* Washington, D.C.: National Academy Press, 1990.

Cook, William J. "The New Frontiers. (NASA's plans to explore Mars and the Earth's moon)" *U.S. News & World Report,* vol. 105, no. 12 (September 26, 1988): 50(2).

Crankshaw, Edward. *Khrushchev: A Career.* New York: The Viking Press, 1966.

"Experience, Analogs and Simulations to Guide Planning for Prolonged Missions" SICSA Outreach. Undated. Publication of the Sasakawa International Center for Space Architecture at the University of Houston's College of Architecture.

Ezell, Edward C., and Linda N. Ezell. *On Mars: Exploration of the Red Planet 1958–1978.* The NASA History Series. NASA SP-4912. Washington, D.C.: NASA, 1984.

Freedman, David H. "Invasion of the Insect Robots." *Discover,* March 1991, pp. 44–51.

Harrison, Albert A. "On Resistance to the Involvement of Personality, Social, and Organizational Psychologists in the U.S. Space Program." *Journal of Social Behavior and Personality,* vol. 1, no. 3 (1986): 315–324.

Harrison, Albert A., and Mary M. Connors. "Crew Systems: Theoretical and Practical Issues in the Fusing of Humans and Technology." Paper presented at the American Group Psychotherapy Association, San Antonio, Tex., February 1991, and at Contact VIII, Phoenix, Ariz., March 1991.

Harrison, Albert A., and Alan C. Elms. "Psychology and the Search for Extraterrestrial Intelligence." *Behavioral Science* 35 (1990): 207–218.

Horio, Paul, et al. "Lunar/Mars Exploration." Pub. 3547-W NEW 1-90. Downey, Calif.: Rockwell International Space Transportation Systems Division, 1990.

Ishikawa, Y., T. Ohkita, and Y. Amemiya. "Mars Habitation 2057: Concept Design of a Mars Settlement in the Year 2057." *Journal of the British Interplanetary Society* 43 (1990): 505–512.

Kanas, Nick. "Psychosocial Support for Cosmonauts." *Aviation, Space, and Environmental Medicine* 62 (April 1991): 353–355.

Kerrod, Robin. *Living in Space.* New York: Crescent Books, 1986.

Lebedev, Valentin. *Diary of a Cosmonaut: 211 Days in Space.* Translated by Luba Diangar. Edited by Daniel Puckett and Dr. C. W. Harrison. Texas: Phyto-Resource Research, Inc., Information Service. 1988 edition.

Ley, Willy, and Wernher von Braun. *The Exploration of Mars.* London: Sidgwick and Jackson, 1956.

"Living in Space: Considerations for Planning Human Habitats Beyond Earth." SICSA Outreach. A publication of the Sasakaw International Center for Space Architecture, at the University of Houston's College of Architecture. Undated.

Lowell, A. Lawrence. *Biography of Percival Lowell.* New York: Macmillan Company, 1935.

Lowell, Percival. *Mars.* Boston: Houghton Mifflin.

McKay, Christopher P. "Exobiology and Future Mars Missions: The Search for Mars' Earliest Biosphere." *Advanced Space Research,* vol. 6, no. 12 (1986): 269–285.

Miles, Frank, and Nicholas Booth. *Race to Mars: The Harper & Row Mars Flight Atlas.* New York: Harper & Row, 1988.

Moore, Patrick. *Guide to Mars.* New York: W. W. Norton & Co, 1977.

———. *The Next Fifty Years in Space.* New York: Taplinger Publishing Co., 1976.

Moore, Patrick, F.R.A.S. "The Planet Mars." *Journal of the British Interplanetary Society,* vol. 14, no. 2 (March–April 1955): 65–84.

Murray, Bruce. *Journey into Space: The First Thirty Years of Space Exploration.* New York: W. W. Norton & Co., 1989.

Myasnikov, V. I. *From "Vostok" to "Mir": Psychological Aspects.* Kosmicheskaya biologiii I aviakosmicheskaya meditsina, No. 6. 1988, pp. 17–23, Moscow, U.S.S.R. Translated by NASA, March 1990.

Noland, David. "Flatland." *Discover,* October 1990, pp. 56–57.

Oberg, Alcestis R. *Spacefarers of the '80s and '90s: The Next Thousand People in Space.* New York: Columbia University Press, 1985.

Oberg, James E. and Alcestis R. Oberg. *Pioneering Space: Living on the Next Frontier.* New York: McGraw-Hill, 1986.

O'Leary, Brian. *Mars 1999: Exclusive Preview of the U.S.-Soviet Manned Mission.* Harrisburg, Pa.: Stackpole Books, 1987.

Paine, Thomas O. "A Timeline for Martian Pioneers." Paper presented at the Case for Mars II Conference, University of Colorado, Boulder, July 10–14, 1984.

———. "Urgently Needed: New Goals for NASA." Testimony before the Subcommittee on Space Science and Applications. U.S. House of Representatives, Washington, D.C., April 5, 1989.

Pryke, Ian. "ESA's Next Decade." *Ad Astra,* May 1989, pp. 16–22.

Richardson, Robert S., and Chesley Bonestell. *Mars.* New York: Harcourt, Brace & World, 1964.

Rockwell International. "Appendix A of Mars Exploration Approach, Task 1.3, Project 22401." Final Report, Technical Volume, September 1990. Proprietary Data Rockwell Corporation.

Rovin, Jeff. *Mars: The First Complete Book of Martian Fact and Fantasy.* Los Angeles: Corwin Books, 1978.

"Space Radiation Health Hazards: Assessing and Mitigating the Risks." SICSA Outreach. Undated. A publication of the Saskawa International Center for Space Architecture at the University of Houston's College of Architecture.

Spangenburg, Ray, and Diane Moser. *Living and Working in Space.* New York: Facts On File, 1989.

Steele, Allen. "The Launch Pad on My Kitchen Table." *Journal Wired,* Summer/Fall 1990, pp. 281–293. Includes a copy of an excerpt from "The Integrated Space Plan, version 1.1" produced by Ronald M. Jones.

Stein, G. Harry. *Rocket Power and Space Flight.* New York: Henry Holt and Co., 1957.

Stoker, Carol, ed. *The Case for Mars III: Strategies for Exploration—General Interest and Overview.* San Diego, Calif.: Univelt Inc., 1989. Proceedings of the Third Case for Mars Conference held July 18–22, 1987, at the University of Colorado, Boulder.

Sullivan, Walter. *Assault on the Unknown: The International Geophysical Year.* New York: McGraw-Hill Book Company, 1961.

The Synthesis Group. "Executive Summary. Report of the Synthesis Group to President Bush." 1991.

Taylor, G. Jeffrey, and Paul D. Spudis. "A Teleoperated Robotic Field Geologist." Proceedings of Space '90 Aerospace/ASCE Albuquerque, N.M., April 22–26, 1990.

Time-Life Books. *Life Search. (Voyage through the Universe).* New York: Time-Life Books, 1988.

———. *The Near Planets. (Voyage through the Universe).* New York: Time-Life Books, 1990.

United States. NASA. "Advanced Missions with Humans in Space." Washington, D.C.: January 8, 1987.

———. "Beyond Earth's Boundaries: Human Exploration of the Solar System in the 21st Century." 1988 Annual Report to the Administrator: Office of Exploration, NASA.

———. "Exploring the Living Universe: A Strategy for Space Life Sciences." Washington, D.C., June 1988.

———. *The Human Factor: Biomedicine in the Manned Space Program to 1980.* Washington, D.C., 1985.

———. *On the Habitability of Mars.* Edited by M. M. Averner and R. D. MacElroy. Prepared by Ames Research Center. Washington, D.C.: NASA, 1976.

———. *Planetary Exploration through the Year 2000, an Augmented Program: Part Two of a Report by the Solar System Exploration Committee of the NASA Advisory Council.* Washington, D.C.: Government Printing Office, 1986.

———. *Report of the 90-Day Study on Human Exploration of the Moon and Mars.* Washington, D.C.: NASA, November 1989.

———. Jet Propulsion Laboratory. *Viking: The Exploration of Mars.* 1984. JPL document 400-219. NASA document EP-208. Washington, D.C.: Government Printing Office, 1984.

———. Scientific and Technical Information Branch. *Life Sciences Space Station Planning Document: A Reference Payload for the Life Sciences Research Facility.* NASA Technical Memorandum 89188. Washington, D.C., NASA, August 1986.

Von Braun, Wernher. *The Mars Project.* Urbana, Ill.: University of Illinois Press, 1962.

Von Braun, Wernher, and Frederick I. Ordway. *Space Travel: A History. An Update of History of Rocketry & Space Travel.* New York: Harper & Row, 1985.

Wilford, John Noble. *Mars Beckons: The Mysteries, the Challenges, the Expectations of Our Next Great Adventure in Space.* New York: Alfred A. Knopf, 1990.

Chapter 4. Mission to Planet Earth

Amsbury, David L. "United States Manned Observations of Earth Before the Space Shuttle." *Geocarto International,* vol. 4, no. 1 (March 1989) : 7ff.

Association of Space Explorers–USA. Annual Report, 1989. San Francisco, Calif.: ASE-USA, 1989.

Bagne, Paul. "Interview: Verner Suomi." *Omni,* vol. 11, no. 10 (July 1989): 60(7).

Begley, Sharon, with Mary Hager. "Feeling the Heat on the Greenhouse." *Newsweek,* May 22, 1989, pp. 78–79.

Booth, William. "Monitoring the Fate of the Forests from Space." *Science* 243 (February 24, 1989): 1428–1429.

Brown, Lester R., et al. *State of the World 1990.* New York: W. W. Norton & Co., 1989.

Clarke, William C. "Managing Planet Earth." *Scientific American,* vol. 261, no. 3 (September 1989): 46–57.

Clery, Daniel. "Sensing Satellites: Who Calls the Tune?" *New Scientist,* 4 May 1991, pp. 17–18.

Cowley, Geoffrey. "Rethinking Evolution." *Newsweek,* October 2, 1989, pp. 38.

Dozier, Jeff. "Looking Ahead to EOS: The Earth Observing System." *Computers in Physics,* May/June 1990, pp. 248–259.

"Earth Observing System: 1989 Reference Handbook." Goddard Space Flight Center: NASA, 1989.

Earth Quest. Quarterly report of the Office of the Interdisciplinary Earth Study of the University Corporation for Atmospheric Research, Boulder, Colorado. Winter 1988, vol. 2, no. 1.

Earth System Sciences Committee, NASA Advisory Council. *Earth System Science: A Program for Global Change.* Washington, D.C.: NASA, May 1986.

Easterbrook, Gregg. "Cleaning Up: Special Report." *Newsweek* July 24, 1989, pp. 26ff.

Fisher, Arthur. "The Wheels within Wheels in the Superkingdom Eucaryotae." *Mosaic,* vol. 20, no. 3 (Fall 1989): 2–13.

Fisk, Lennard A. "Observing the Earth from Space: Orbital Studies Can Help Protect the Global Environment and Preserve Our Economic Well-being." *The Space Times,* January–February 1991, pp. 9–10ff.

Furuta, H. "Earth Observation in Japan." Notes of speech given at the ISY Mission to Planet Earth Conference, Durham, New Hampshire, April 29–30, May 1, 1989.

Gavaghan, Helen. "An Expedition to Earth." *New Scientist,* 29 July 1989, pp. 26–27.

Grove, Noel. "Air: An Atmosphere of Uncertainty." *National Geographic,* vol. 171, no. 4 (April 1987): 502–537.

Hansen, James E. "The Greenhouse Effect: Impacts on Current Global Temperature and Regional Heat Waves." Statement presented to the United States House of Representatives, Committee on Energy and Commerce, Subcommittee on Energy and Power, July 7, 1988.

Hecht, Susanna, and Alexander Cockburn. "Brazil on Fire." *LA Weekly,* December 8–14, 1989, pp. 22ff.

Hecht, Susanna, and Alexander Cockburn. *The Fate of the Forest: Developers, Destroyers and Defenders of the Amazon.* New York: Harper Perennial, 1990.

Helfert, Michael R. "NASA Human-Directed Observations of the Earth from Space: An Introduction." *Geocarto International,* vol. 4, no. 1 (March 1989): 3ff.

Helfert, Michael R., and Charles A. Wood. "The NASA Space Shuttle Earth Observations Office." *Geocarto International,* vol. 4, no. 1 (March 1989): 15ff.

Isbell, Douglas, "NASA, White House Name Panel for EOS Examination." *Space News,* April 15–21, 1991, pp. 17.

Jordan, Carl F., ed. *Amazonian Rain Forests: Ecosystem Disturbance and Recovery.* New York: Springer-Verlag, 1987.

Kaiser, Charles. *1968 in America: Music, Politics, Chaos, Counterculture, and the Shaping of a Generation.* New York: Weidenfeld and Nicolson, 1988.

Kelley, Kevin W. *The Home Planet.* Reading, Mass.: Addison-Wesley Publishing Co., 1988.

Lowenstein, Frank. "Seasons, Seas, and Satellites" *Air & Space,* February/March 1988, pp. 42ff.

Lulla, Kamlesh P., et al. "Earth Observations during Space Shuttle Flight STS-41: Discovery's Mission to Planet Earth." *Geocarto International,* vol. 6, no. 1. (1991): 69ff.

Margulis, Lynn, and Dorion Sagan. *Microcosmos.* New York: Simon and Schuster, 1986.

Matthews, Samuel W. "Under the Sun: Is Our World Warming?" *National Geographic,* vol. 178, no. 4 (October 1990): 66–99.

Myers, Dr. Norman, ed. *Gaia: An Atlas of Planet Management.* Garden City, N.Y.: Anchor Press/Doubleday & Company, 1984.

Nilsson, Sten, and Peter Duinker. "The Extent of Forest Decline in Europe: A Synthesis of Survey Results." *Environment,* vol. 29, no. 9 (November 1987): 4ff.

"Overview of ISCCP." March 1989. Discussion of the International Satellite Cloud Climatology Project (ISCCP) established as part of the World Climate Research Program.

Ramanathan, V., Bruce R. Barkstrom, and Edwin F. Harrison. "Climate and the Earth's Radiation Budget." *Physics Today,* May 1989, pp. 22ff.

Ramanathan, V., and W. Collins. "Thermodynamic Regulation of Ocean Warming by Cirrus Clouds Deduced from Observations of the 1987 El Niño." *Nature,* vol. 351 (May 2, 1991): 27ff.

Ramanathan, V. et al. "Cloud-radiative Forcing and Climate: Results from the Earth Radiation Budget Experiment." *Science* 243 (6 January 1989): 242–243.

Raval, A., and V. Ramanathan. "Observation Determination of the Greenhouse Effect." *Nature* 342 (14 December 1989): 758ff.

"Rediscovering Planet Earth." *U.S. News & World Report* (October 31, 1988): 56ff.

Ruckelshaus, William D. "Toward a Sustainable World." *Scientific American,* vol. 261, no. 3 (September 1989): 166–175.

Sagan, Dorion, and Lynn Margulis. "Gaia and the Evolution of Machines." *Whole Earth Review,* Summer 1987, pp. 15–21.

"The Space Explorer." Newsletter of the Association of Space Explorers. January 1991 issue.

Stewart, Doug. "Eyes in Orbit Keep Tabs on the World in Unexpected Ways." *Smithsonian,* vol. 19, no. 9 (December 1988): 70(10).

Testimony of Dr. Barrett N. Rock before the Subcommittee on Natural Resources, Agricultural Research, and Environment, Committee on Science, Space, and Technology, U.S. House of Representatives.

"Time Essay: Of Revolution and the Moon." *Time,* January 3, 1969 p. 17.

Time Special Issue: Planet of the Year. *Time,* vol. 133, no. 1 (January 2, 1989): 3ff.

White, Frank. *The Overview Effect: Space Exploration and Human Evolution.* Boston: Houghton Mifflin, 1987.

Chapter 5. Celestial Sentinels

Adams, Richard E. W. *Prehistoric Mesoamerica: Revised edition.* Norman, Okla.: University of Oklahoma Press, 1991.

Adams, R.E.W., W. E. Brown, Jr., and T. Patrick Culbert. "Radar Mapping, Archeology, and Ancient Maya Land Use." *Science,* vol. 213, no. 4515 (25 September 1981): 1157.

Bailey, Lt. Col. Rosanne, and Lt. Col. Thomas Kearney. "Combat Enters Hyperwar Era." *Defense News* (July 22, 1991).

Berman, Bob. "Satellite Season." *Discover,* May 1991, pp. 32–33.

Bluth, B. J. "A Critical Look at Space Technology and the Developing World." *Advanced Space Research,* vol. 3, no. 7 (1983): 13–22.

Bornet, Vaughn Davis. *The Presidency of Lyndon B. Johnson.* Lawrence, Kans.: University Press of Kansas, 1983.

Broad, William J. "What's Next for 'Star Wars'? 'Brilliant Pebbles.' " *The New York Times,* April 25, 1989, pp. 19ff.

Buenneke, Richard H., Jr. "Everything You Always Wanted to Know About Military Space Programs But Didn't Have the Security Clearance to Ask." *Final Frontier,* November/December 1990, pp. 32–36.

Burrows, William E. *Deep Black.* New York: Random House, 1986.

Chetty, P.R.K. *Satellite Technology and Its Applications.* Second edition. Blue Ridge Summit, Pa.: Tab Books, 1991.

Clarke, Arthur C. *Ascent to Orbit: A Scientific Autobiography.* New York: John Wiley & Sons, 1984.

Clarke, Arthur C. "Extra-terrestrial Relays: Can Rocket Stations Give World-wide Radio Coverage?" *Wireless World,* October 1945, pp. 305–309.

Covault, Craig. "Major Space Effort Mobilized to Blunt Environmental Threat."

Part of Segment: Mission to Planet Earth. *Aviation Week and Space Technology,* March 13, 1989, pp. 34ff.

Dalgleish, D. I. *An Introduction to Satellite Communications.* London: Peter Peregrinus Ltd., 1989.

Deuel, Leo. *Flights into Yesterday: The Story of Aerial Archaeology.* London: Macdonald, 1971.

Drury, S. A. *A Guide to Remote Sensing: Interpreting Images of the Earth.* Oxford: Oxford University Press, 1990.

Franke, Richard W., and Barbara H. Chasin. *Seeds of Famine: Ecological Destruction and the Development Dilemma in the West African Sahel.* Totowa, N.J.: Allanheld, Osmun & Co, Publishers, Inc., 1980.

Gallenkamp, Charles. *Maya: The Riddle and Rediscovery of a Lost Civilization.* New York: The Viking Press, 1985.

Glatzer, Hal. *The Birds of Babel: Satellites for the Human World.* Indianapolis, Ind.: Howard W. Sams & Co., Inc., 1983.

Fowler, Martin J. F. "Satellite Archaeology." *Spaceflight* 33 (August 1991): 281.

Gilmartin, Patricia. "Gulf War Rekindles U.S. Debate on Protecting Space System Data." *Aviation Week and Space Technology,* April 29, 1991, p. 55.

Kahn, Herman. *Thinking about the Unthinkable.* New York: Horizon Press, 1962.

Kingwell, Jeff. "The Militarization of Space: A Policy Out of Step with World Events?" *Space Policy,* May 1990, pp. 107–111.

Lafeber, Walter. *America, Russia, and the Cold War.* New York: John Wiley and Sons, 1967.

McLucas, John L. *Space Commerce.* Cambridge, Mass.: Harvard University Press, 1991.

McLuhan, Marshall, and Bruce R. Powers. *The Global Village: Transformations in World Life and Media in the 21st Century.* New York: Oxford University Press, 1989.

Mitchell, Sally, ed. *Victorian Britain: An Encyclopedia.* New York: Garland Publishing, Inc., 1988.

O'Neill, E. F., ed. *A History of Engineering and Science in the Bell System: Transmission Technology (1925–1975).* Prepared by Members of the Technical Staff, AT&T Bell Laboratories. Indianapolis, Ind.: AT&T Bell Laboratories, 1985.

Osman, Tony. *Space History.* New York: St. Martin's Press, 1983.

Peebles, Curtis. *Guardians: Strategic Reconnaissance Satellites.* Novato, Calif.: Presidio Press, 1987.

Pierce, John R. "The Telephone and Society." In Pool, Ithiel de Sola. *The Social Impact of the Telephone.* Cambridge, Mass.: The MIT Press, 1977.

Robinson, Howard. *The British Post Office: A History.* Princeton, N.J.: Princeton University Press, 1948.

Stewart, Doug. "Eyes in Orbit Keep Tabs on the World in Unexpected Ways." *Smithsonian,* December 1988, pp. 70–81.

Szulc, Tad. *Then and Now: How the World Has Changed Since WWII.* New York: William Morrow and Company, 1990.

Trux, Jon. "Desert Storm: A Space-age War." *New Scientist,* 27 July 1991, pp. 30–34.

U.S. Department of Defense. *Defense 88: Looking at Space Past Present and Future.* Alexandria, Va.: American Forces Information Service, 1988.

Walker, Susan. "Storms Ahead for U.S. Weather Satellites." *New Scientist,* 27 July 1991, p. 22.

Chapter 6. New Frontiers

Cameron, A.G.W. "Giant Impact Theory of the Origin of the Moon." *Planetary Geosciences—1988.* Washington, D.C.: NASA Office of Management, Scientific and Technical Information Division, 1989.

Clark, Phillip. *The Soviet Manned Space Program.* New York: Orion Books, 1988.

Clarke, Arthur C. *2001: A Space Odyssey.* New York: New American Library, 1968.

Covault, Craig. "NASA Accelerates Lunar Base, Mars Studies for Input to New Administration." *Aviation Week and Space Technology,* vol. 129, no. 22 (November 28, 1988): 45(2).

Harding, M. Esther. *Woman's Mysteries: Ancient and Modern.* London: Longmans, Green & Co., 1935.

Hoverstein, Paul. "Back to the Moon." *Discover,* vol. 11, no. 9 (September 1990): 24(1).

Mendell. W. W., ed. *Lunar Bases and Space Activities of the 21st Century.* Houston, Tex.: Lunar and Planetary Institute, 1985.

Moore, Patrick. *The Moon.* The Rand McNally Library of Astronomical Atlases for Amateur and Professional Observers. New York: Rand McNally & Co, 1981.

O'Neill, Gerard. "The New American Frontier; The Shuttle Has Put Us in a Position to Prosper in Space—If We Apply Old-fashioned Pioneer Spirit and Sage Use of Resources to the Task." Special Report. *Discover* 6 (November 1985): 70(8).

Sheehan, William. *Planets and Perception: Telescopic Views and Interpretations: 1609–1909.* Tucson: University of Arizona Press, 1988.

Time-Life Books. *Spacefarers.* Part of the series *Voyage through the Universe.* Richmond, Va.: Time-Life Books, 1990.

Urey, Harold C. *The Planets: Their Origin and Development.* New Haven: Yale University Press, 1952.

Von Braun, Wernher. *First Men to the Moon.* New York: Holt, Rinehart and Winston, 1960.

Wolfe, Tom. *The Right Stuff.* New York: Farrar, Straus, Giroux, 1979.

Epilogue. The Shape of Things to Come

Forward, Robert. "Advanced Space Propulsion." *The Journal of Social, Political and Economic Studies,* vol. 15, no. 4 (Winter 1990): 387(16).

McKay, Christopher P., and Robert H. Haynes. "Should We Implant Life on Mars?" *Scientific American,* vol. 263, no. 6 (December 1990): 144(1).

Credits

Pps. i–vii: Images of Venus courtesy of NASA; pps. 2–3: Ordway Collection/U.S. Space and Rocket Center; p. 4: K. E. Tsiolkovsky Museum for Cosmonautics, Kaluga, Russia; p. 5: K. E. Tsiolkovsky Museum for Cosmonautics, Kaluga, Russia; p. 6: K. E. Tsiolkovsky Museum for Cosmonautics, Kaluga, Russia; p. 7: K. E. Tsiolkovsky Museum for Cosmonautics, Kaluga, Russia; p. 8: Mary Evans Picture Library; p. 9: Mary Evans Picture Library; p. 10, right: Mary Evans Picture Library; p. 10, left: K. E. Tsiolkovsky Museum for Cosmonautics, Kaluga, Russia; p. 11: Mary Evans Picture Library; p. 12, top: Sovfoto; p. 12, bottom right: K. E. Tsiolkovsky Museum for Cosmonautics, Kaluga, Russia; p. 12, bottom left: Warren Morgan; p. 14, top: Scala/Art Resource, N.Y.; p. 14, bottom: Giraudon/ Art Resource, N.Y.; p. 15: Trustees of the British Museum; p. 16, top right: Giraudon/Art Resource, N.Y.; p. 16, bottom right: Snark/Art Resource, N.Y.; p. 16, top left: courtesy of the National Portrait Gallery; p. 21: Goddard Collection, Clark University, Worcester, Massachusetts; p. 24: Goddard Collection, Clark University, Worcester, Massachusetts; p. 30: Goddard Collection, Clark University, Worcester, Massachusetts; p. 31: courtesy of the Hermann Oberth Museum; p. 34: courtesy of the Hermann Oberth Museum; p. 38: AP/Wide World Photos; p. 40: the Kobal Collection; p. 42: Goddard Collection, Clark University, Worcester, Massachusetts; p. 45: Ordway Collection/U.S. Space and Rocket Center; p. 46: Ordway Collection/U.S. Space and Rocket Center; p. 47, bottom: courtesy of the Hermann Oberth Museum; p. 47, top: Ordway Collection/U.S. Space and Rocket Center; pps. 48–49: NASA/JPL art by Ken Hodges; p. 50: Ordway Collection/U.S. Space and Rocket Center; p. 51: Mitchell R. Sharpe; p. 55: Mitchell R. Sharpe; p. 58: U.S. Space and Rocket Center; p. 61: U.S. Space and Rocket Center; p. 65: Trustees of the Imperial War Museum, London; p. 66: Trustees of the Imperial War Museum, London; p. 68: Sovfoto; p. 73: Sovfoto; p. 77: Ordway Collection/U.S. Space and Rocket Center; p. 81: Sovfoto; p. 83: courtesy of NASA; p. 85: courtesy of NASA; p. 87: courtesy of NASA; p. 88: Sovfoto; p. 89, left: LBJ Library; p. 89, right: courtesy of NASA; p. 94: courtesy of NASA; p. 98: courtesy of NASA; p. 100: courtesy of NASA; p. 101: courtesy of NASA; p. 102: NASA art by Pat Rawlings/SAIC; p. 104: NASA art by Pat Rawlings/SAIC; p. 105: courtesy of NASA; p. 107: NASA/JPL art by Ken Hodges; p. 108: artwork by Carter Emmart; p. 109, top: World Space Foundation/Carol R. Stoker; p. 109, bottom: World Space Foundation/Richard Dowling/Metavision; p. 110: McDonnell Douglas Photos; pps. 112–113: QED Science Effects Unit; p. 114: Lowell Observatory Photograph; p. 115: Ordway Collection/ U.S. Space and Rocket Center; p. 117, top: courtesy of the University of Arizona Press; p. 117, bottom: Lowell Observatory Photograph; p. 118: AP/Wide World Photos; p. 119: Lowell Observatory Photograph; p. 120: Mary Evans Picture Library; p. 121: Lowell Observatory photograph taken by L. J. Martin at Mauna Kea Observatory. Color compositing and processing courtesy of the U.S. Geological Survey, Branch of Astrogeology. 1988 observations supported by The National Geographic Society; p. 123: courtesy of NASA; p. 125, top: courtesy of NASA; p. 125, bottom: U.S. Geological Survey, Flagstaff; p. 127, top: photographs available from the U.S. Department of the Interior, U.S. Geological Survey, EROS Data Center. Landsat E-1039-18143-5, 31 August 1972; p. 127, center: photo by V. Baker, University of Arizona; p. 127, bottom: U.S. Geological Survey, Flagstaff; p. 129: courtesy of NASA; p. 133: National Science Foundation; p. 135: NASA photo by Robie Vestal; p. 136: NASA photo by Dale Andersen; p. 139: WQED Science Effects Unit, Don Davis, Tony Meininger; p. 141: courtesy of the artist Robert McCall; p. 142: Sovfoto; p. 143: Sovfoto; p. 146: Space Commerce Corporation; p. 147: Ordway Collection/U.S. Space and Rocket Center; p. 152: NASA artwork by Pat Rawlings/SAIC; p. 153: courtesy of Martin Marietta Astronautics Group; p. 154: courtesy of Martin Marietta Astronautics Group; p. 155: QED Science Effects Unit; p.

Index

A

A-1 rocket, 59
A-2 rockets, 59
A-3 rockets, 60
A-4 rockets. *See* V-2 rockets
Abbot, Charles, 35
Accidents, 46
Acid-rain, 190, 192
Aerobrake, **158, 160**
Agrhymet satellite station, **251**
Aldrin, Buzz, **255,** 256
Aliens. *See* Extra-terrestrial life
Alvarez, Luis, 216
Amazing Stories, **120,** 122, 216
Amazon rain forest, 194–200, **195**
Ambler, 132
American Meteorological Society, 244
American Rocket Society, 216
Ames Research Center, 136
"Analysis of the Potential of an
    Unconventional Reconnaissance
    Method," 219
Andorfer, Gregory, xiv
Animals in space
    dogs, 74–75
    Russian, Laika, 84
Antarctica, 137
    fossil traces on, **133**
    Lake Vanda, **135**
    microbial mats, **135**
    studies in, on problems of isolation and
        stress, 152, 154
    yeast in rocks, 165
Apollo program, 88–90, 93, 95, 257–58
    Apollo 8, 185
    Apollo 11, 96, 97, 255, 257
        lander, **94**
    Apollo 17, 255, 257
Apollo-Soyuz mission, 97
Arecibo Observatory, 276
Ares launch vehicle, **154**
Argon, 106
Ariane 4 rocket, 107, **289**

Aristotle, 273, 311
Armstrong, Neil, 100, 255, 256
Asimov, Isaac, 122, 257
Astounding Science Fiction, 216
Astrology, 14
Astronauts
    profile of ideal, 152, 154
    Mercury, **87**
Astronomy, 15, 120
Atlas rocket, 295
Atomic bomb
    first detonated at Alamagordo, 71
    first one exploded by Soviets, 74
    Manhattan Project, 71–74

B

Babassu palm, 197
Babylonian clay tablet, **175**
Bacteria. *See* Microbes
Baikonur Cosmodrome, 18
Bailey, Col. Rosanne, 231
Bardeen, John, 234
Barnard, E.E., 121
Barsoom, 121–22
Bartoe, John-David, 179
Becker, Karl, 55, 56
Bell Labs, 234, 236
Berezovoy, Anatoli, 150
Berkner, Lloyd, 78
Bermuller, Hans, **47**
Big Bang theory, 276, 291
Big Bird, 220
Biosphere Earth, 157, 159, 207–9
    idea developed by Vernadsky, 179, 182
Black holes, 276, 311
"Black triangle," 189–94, 192, 193, **194**
    Landsat map of, **191**
Blake, William, 4
"Blitzkrieg," 228, 230
Bonestell, Chesley, 147, 263, 280
Books on space. *See* Science fiction: books
Bradbury, Ray, 114, 122, 161

Brattain, Walter, 234
Braun, Magnus von, **51,** 53
Braun, Wernher von, 43, 47, 50–71, **51, 85,**
    138, 219, 234
    arrested, in 1944, 70–71
    biographical background, 53, 54
    builds rocket for man-on-the-moon project,
        91
    concept of space station, **147**
    controversy over his knowledge of slave
        labor at Nordhausen, 67
    hired by Dornberger to head up rocket
        artillery unit, 56
    launches *Explorer I,* 84–85
    leaves NASA, 97
    plans space station, 87
    and problems of weightlessness, 146
    and *Saturn V,* 96
    surrenders to U.S. troops, 52, 53
    work at White Sands, 75–76, 78–79
Brazilian Institute for Space Research (INPE),
    198, 200
Bretz, Harlen, 126, 128
Brilliant Pebbles, **221, 223**
British Interplanetary Society, 260
Bronowski, Jacob, 13
Brooks, Rodney, 134
Burroughs, Edgar Rice, 121–22, 130, 161
Bush, George, 140
Bussard, R.W., 306, 307
Buzz bomb, 51n

C

Cable television systems, 239–42
"Canadarm," **264**
Canali, 116, 117, 118
Caranthus roseus, 196–97
Carbon sticks, 46
Carcinogens, 185, 189–90
Carnegie Institute, 311
Carpenter, Scott, **87**
Cassiopeia, **280**

Cave paintings, Lascaux, **177**
Centaur rocket, 295
Cernan, Eugene, 258
Challenger, 99–100, **100**
Cheyenne Mountain, 222, 224
Chibis Vacuum Suit, **146**
Childhood's End, 216
Chinese Long March rockets, 200
Chlorophyll, 247
    concentrations, in ocean, **208–9**
Chryse Plantatia, 300–301
Churchill, Winston, 231–32
CIS (Commonwealth of Independent States)
    space programs, 17–19
Clarke, Arthur C., xiv, 67, 122, **232**, 234, 240,
    250
    first proposed communications satellites,
        216, 217, 218, 224, 232
    suggests mining the moon, 268, 286
"Clipper Ships of Space," 106
Cloud cover, **201**
Clovis, The, **12**
COBE (cosmic background explorer), 17
Collier's magazine, 77, **115**, 139, 232, 260,
    261
Collins, Claude R., 28
Collins, Michael, 144
Columbia space shuttle, 97, 228
Computers, 202, 203
    of the future, 309
Conquest of Space, The, 216
"Constant source" program, 231
Controlled-Environment Life Support System
    (CELSS), **156,** 157
Cooper, Gordon, **87**
Copernican model of the universe, **16**
Corpus Hermeticum, 14
Cosmic Background Explorer (COBE), **291,**
    311
Cosmonauts, 6, 88, 90, 91
    first woman, 91
    psychological problems of, in microgravity,
        148–52
Cosmotheoros, 303
Cretaceous period, **196**

D

Darwin, Charles, 183, 185, 194, 310
Debris
    left on the Moon, 257
    orbital, around Earth, **214,** 276
Defense News, 231
Defense Support Program (DSP)
    satellite, **210–11, 229,** 230
Delta rocket, 236
Delta II, **215**
Der Vorstoss in den Weltenraum (The
    Advance into Space), 36, 37
Die Frau Im Mond (The Woman in the
    Moon), 40, 41, 46–47, 54, 55
    actors in, **2–3,** 46
    poster from, **40**
    rocket in, **45**
Die Rakete zu den Planeträumen (By
    Rocket to Planetary Space), 33–35, 36

Diogenes, Antonius, 6
Discovery, 274
Disturbing the Universe, 304
DNA, 237, **238, 239**
Donati's comet, 116, 118
Dornberger, Walter, 51, 52, **55,** 55–56, 59, 60,
    70
    put in charge of V-2 project, 55
Drake, Frank, 309
Dreams of Earth and Sky, 303, 307
Dressler, Alan, 311
Durand, Asher B., 175
Dyson, Freeman, 299, 304–5

E

Early Bird, 237
Early man, 13, 15
Earth, The, xii, 174–209
    atmosphere on, 182, 183
    the "black triangle," 189–90
    composite satellite image of, **176**
Earth-observing instruments, 204–207
    ecological damage, 185, 189
        destruction of forests by acid-rain, 190,
            192
        exploitation of resources, 178
        extinction of animal species, 185
        pollution of air and water, 190
        slash-and-burn expansion policy, Brazil,
            196
    economic and social links with the Moon,
        284
    effects of Industrial Revolution upon, 178
    evolution of life-forms on, 184–85
    Global Biosphere, 207, 208–9
    global climate, 200
    global warming, 200, 201
    Life, importance of preserving it, 196
    Lovelock's theories, 182, 183
    microbes, first forms of life on, 184
    object collides with, 252–53, **256**
    as one integrated system, 179
    photographs, by astronauts in orbit, 189
    populations, 177
    saving the planet, 207
    surface temperatures, **203**
    Vernadsky's theories, 179, 182, 183
    as viewed from *Apollo 8,* 175
    viewed from lunar orbit, **172–73, 197**
Earth Observing System (EOS), 204, 205, 206,
    207
Earth Resources Satellite (ERS), **205**
Echo, 87, 234, **235,** 236
Edison, Thomas, 5
Ehricke, Krafft, 295, 296, 297
Einstein, Albert, 72, 185
Eisenhower, Dwight D., 82
Electronic monitoring technologies, 218–19
    See also Satellites: types: spy
El Niño, 200
End of a Two-Week Long Lunar Day, 278–79
Energia rocket, 290
Energia Scientific Industrial Company, 18–19
Entropy, law of, 196
Eratosthenes, 177

European Centre for Nuclear Research
    (CERN), 306
European Space Agency (ESA), 18, 290, 297
Evolution, theory of, 194, 295–96
Exercise, in space, 144, 145, 272
Explorer I, 85, **85,** 86
Explorers, early world, 19
Extended Launch Vehicle, 257, **257**
Extra-terrestrial life, 128, 130–32, 303–9
"Extra-Terrestrial Relays: Can Rocket Stations
    Give World-Wide Radio Coverage?",
    232

F

Factory, steam powered, 178
"Father of Cosmonautics." *See* Tsiolkovsky,
    Konstantin
"Fat Man," 71, 72
    See also Atomic bomb
"Ferret," 224
Fi-103, 51n
Flammarion, Camille, 116, **118**
Food, growing in space, 156–57, 159, 161
Folsome, Clair, 157n
Forbidden Planet, 146
Forests. *See* Rain forests
Fornax A, Radio Galaxy, **280**
Forward, Robert L., 305, 307
Freedom space station, 111, 204
Freeman, Fred, 232
"Free Space"
    sketch of ship, by Tsiolkovsky, **10**
Friedman, Herb, xi
Friedman, Imre, 137, 138, 165, 168
Friedman, William F.
    attempts to detect alien transmissions, 115–
        16, 282
From the Earth to the Moon, 6n, 33
    illustrations from, **8, 9, 10, 11**
Fuels, 46, 59
Funding
    for Apollo project, 90
    budget problems at NASA, 96–97, 99, 128
    for research, 25
    for satellite system, 286
    for *Space Age* tv series, xiv
    for space program, 220
Fusion reactor, 288
    JET, 288n
"Future of World Communications, The," 216,
    218

G

Gagarin, Yuri, 6, 90
    first orbits the earth, 88, **88**
Gaia Hypothesis, 183
Galaxies, **280,** 311
Gamma Ray Observatory, **275,** 276, **281**
Gamma ray radiation, 281
Gemini program, 83
Geochemistry, 179
Geo-positioning satellites, 249–50
Geostationary satellite, 218n, 232

Germany
  early interest in rocketry, 35, 36, 37
  weapon development, 27, 51–67
Gernsback, Hugo, 122
GIRD X. GIRD (09) rocket, **68**, 69
Glenn, John H., Jr., **87**
  first U.S. orbital flight, 91
Global Biosphere, 207, **208–9**
Global climate, 200
Global Outposts, 266
Global Positioning Satellite system
  during the Gulf War, **215**
Global Positioning System (GPS), 249–50
"Global village," 237
Global warming, 200, 201
  graph of, since 1880, **204**
Global weather patterns, 202
Glushko, Valentin, 95
Goddard, Robert H., 19–31, **21**, 35, 76, 102,
    103, 202
  biographical background, 22
  education, 23
  grant from Guggenheims, 57
  ideas on rocketry and space travel, 22–23,
    24
  investigates use of rockets as weapons, 26–
    27
  launches first liquid fuel rocket, 41, **42**
  publishes paper on rocketry work, 1920, 27
  receives grant from Smithsonian, 25
  sketches from his notebook, **30**
  use of moon materials for space travel, 268
  view of the far future, 309
  wins first patent, 23
Goddard Institute for Space Studies, 204
Göring, Hermann, 59
GPS World, 249
Greczynski, Jan, **190**
Greenhouse effect, 202
Grissom, Virgil I., **87**
Grötrupp, Helmuth, 70, 71
Ground Controlled Approach Radar, 216
Group for the Study of Rocket Propulsion
    Systems (GIRD), **68**, 69, 70
Groves, Leslie R., 72
Guidestar Catalogue, 274
Gulf War, **215**
  uses of satellites during, 214, 231, 232
Gyroscopic air vanes, 57

H

Hahn, Otto, 72
Halley's comet, 107, 292
Hansen, James, 204
Haskins, Lawrence, 268, 272
Heavy Lift Launch Vehicles, 263
Hebes Chasma, 163, 165
Heinisch, Kurt, **47**
Heinlein, Robert, 122
Helfert, Mike, 185, **185**, 189
Heliogyro project, **48–49**, 106, **107**
Helium 3, 286, 288
Henry, O., 148
Hermes, 290
Himmler, Heinrich, 71

Hitler, Adolf, 54, 59
  V-2 rocket demonstrated for, 62–64
Hodgkins, Thomas G., 25
Homo habilis, 177
HOPE, 290, 292
House Committee on Science and
    Astronautics, 84
Hubble Space Telescope, **272**, 273, **274**, 276,
    281, 311
Hughes Research Laboratories, 305
Human communications
  advances in, 237
Human expeditions. *See* Manned exploration
Hurricanes
  Gilbert, 244
  Hugo, 242, **242**, 244
Huygens, Christiaan, 6, 303
Hydroponic crops, **156**, 157
Hyperwar, 231

I

IBAMA, 198
ICBMs, 72, **73**, 74, 79
Ilmenite, 265
Industrial Revolution, 178
Industrial Space Facility, 297
Inertial guidance system
  developed at Peenemünde, 60, 62
Infrared radiation, 305
Infrared telescopes, 281
Ino engine, 103, 106
Institute for Space and Astronautical Science
    (ISAS), 292
Institute of Soil Science and Photosynthesis,
    165
Intelsat, 236–37
International Geophysical Year of 1957–58, xi,
    78
International satellite communication, 236–37
International space station, **262**
International Space Year, 1992, xi
Interplanetary travel
  requirements for, 140, 142
Interstellar ramjet, 306–7
Interstellar travel, 305–7
"Investigation of World Spaces with Reaction
    Machines," 6
  Tsiolkovsky's reaction machine, 7
Ion engine, 103, 106
Ionians, 15
Ion propulsion, 29, 102
Iron, 265

J

Japan
  space program, 292
JATO project, 86
JET, 288n
Jet Propulsion Laboratory (JPL), xii, 85–86, 87,
    106, 107, 132, 134
Johnson, Lyndon, 82, 220, 222
Johnson Space Center, 268
Jornada del Muerto, 71

Journey to the Center of the Earth, 6n
Jupiter-C, 85, 86

K

Kaluga, 4
Kammler, Hans, 51
Karman, Theodore von, 85, 86
Kearney, Lt. Col. Thomas, 231
Kennedy, John F., 88–89, **89**
  memo to Johnson, **89**
Keyes, Roger, 232
Kompfner, R.W., 234, 236
Korolev, Sergei Pavlovich, 67, **68**, 69–70, 74,
    75, 91, 165, 234
  conceives Soyuz program, 92, 93
  death of, 93
  develops first satellite, 78–79
  first successful test-firing of ICBMs, 79
  launches dog-carrying satellite, 82
  launches *Luna* 2 and 3, 86–87
  launches *Sputnik I,* 80–82
  sends first man into space, 88
Krushchev, Nikita, 70, 75, 82, **88**, 92, 93
Krušnéhory Mountains, **193**
Krypton, 102, 106
Kubrick, Stanley, 257

L

LaCrosse satellite, 224
Lagoa Dos Patos lagoon, Brazil, **188**
Lake Chad, 186, **186**
Lake Hoare, **136**, 137, 138
Landsat satellite, 126–28, 190–92, 247
  images, used to assess acid-rain, 190,
    191
  photographs of Scablands, **127**
Land vegetation patterns, 207, **208–9**
Lang, Fritz, 40–41, 297
Lasser, David, 44, 216
"Leadership and America's Future in Space,"
    140
Lebedev, Valentin, 150
Lederberg, Joshua, 165
Lenin, Vladimir Ilyich, 37
Le Planete Mars, 118
Ley, Willie, 44, 46, 47, 54
Life forms in space
  search for, 128, 130–32
Lindbergh, Charles, 57
Lives of a Cell, The, 308
Lockheed Missiles Systems, 219, 220
Locust swarms, **248**
London
  bomb shelters in, **66**
  first V-2 attacks on, 64, **65**
Long Playing Rocket satellite, 234
Los Alamos, 72
Lovelock, James, 182
Lowell, Percival, **114**, 116–22, 161, 307
  views Mars from his observatory, 118–
    21
Lunar Base Studies, 263
Lunar lander, **102**

Lunar soil and rock, 255, 258, **258**
  composition of, 268
Lunar telescope, 270–71

M

McCandless, Bruce, **101**
McIDAS satellite system
  composite global view of Earth's weather
    systems, **244**
McKay, Chris, **136**, 137, 138, 163, 165, 168, 310
McLuhan, Marshall, 237, 251
Macroscope, 190
Magdalenians, 177
Malina, Frank J., 85, 86
Manned exploration, xiii. *See also* Apollo
    program; Space shuttle
  to Mars, proposal for, 138
  moon landings
    Apollo 11, 96–97
    Mercury, **87**, 88, 142
    U.S. proposes, 89–90
Manned Maneuvering Unit (MMU), **101**
Man Tended Free Flier, 297
Maps, 177
  Babylonian, of the world, 175, **175**, 177
Marconi, Guglielmo, 5, 115
Margulis, Lynn, 171, **183**, 183–85, 310
Mariner, 97, 122–28
  first images from, **123**
Mars, 100, 103, 106, 114–71, 120, **170**, 196
  atmosphere on, 130, 182
  cargo ship orbiting, **302**
  Chryse Planitia, **104**
  composite photo of, **168**
  early beliefs about, 120–23
  earthlings on, **303**
  first images of, from *Mariner 4,* **123**
  first photograph of, by *Viking I,* **129**
  future spaceship to, 112–13, **139**, **161**
  globe of, created by Lowell, **119**
  life-theories, 165
  making it habitable, 171
  manned expeditions to, proposal, 138, 272
  manned exploration of the surface of, 161,
    163, 165, 260
  maps of, **117**
  Martians, Collier's magazine, **115**
  Observer spacecraft, 163
  Olympus Mons, **125**, 126
  photograph, frost on Martian landscape,
    129
  photograph from Earth, **121**
  planetary metamorphosis of, 309–10
  plans to colonize, 168–69, 171
  report of National Commission on Space,
    140
  robotic exploration of, methods, 132, 134
  rover, **162**
  sunset on, from *Viking I,* 300–301
  temperature on, 130
  testing for life-forms on, 130–32
  training for a trip to, 154
  transfer vehicle, **152**, **153**
  transportation depot in orbit, **141**
  Valles Marineris, **125**, 126

Mars Direct, **154**
"Mars Project, The," 138
Martin Marietta Astronautics Group, 154
Mass driver, on the Moon, **285**, 286
Matsunaga, Spark, xi–xii
Max, 59
Mendeleyev, Dmitri, 26
Mendell, Wendell, 263, 265, 295
Men in the Moon, The, 299
Mercury program, 87, 88, 142
Meredith, George, 254
Meteostat, 247
"Method of Reaching Extreme Altitudes, A,"
  25, 76
Metropolis, 40
Michelson, A. A., 27
Microbes
  early life-forms on Earth, 184
  found on Antarctica, 137–38
  for "terraforming" Mars, 310
Microgravity, 272
  psychological problems, of crews in, 148–
    52
  research on effects of, 142, 144, 145, 146
MiG 29, **220**
Military satellites. *See* Satellites: types: spy
Milky Way, 307, 311
MIR, **143**, 297
  MIR Quantum-Soyuz TM-3, **143**
  MIR Salyut-Cosmos, **143**
  MIR space station, 18
MIRAK rockets, 43, 54
Missile Defense Alarm System (MIDAS), 222
Missiles
  ICBM, 72, **73**, 74
  R-7 booster, 74
  T-2, 74
Mission to Mars, 144
Mission to Planet Earth, 206–7
Mitsubishi Corporation, 286, 292
Mittelwerk, 64n
Model B rockets, **34**, 35
Model E rockets, 35
Modular launch vehicle, **289**
Moon, 252–301, 286
  as an astronomical observatory, 273
  biospheres on, 299
  colonizing, 296
  debris left on, 257
  developing a lunar economy, 295–96
  early beliefs about, 256–57
  economic and social links with Earth, 284
  export and marketing of moon products,
    296–97, 299
  first earthlings to walk on, 256
  fuel from, 265
  future plans for exploration of, 260, 263
  gravitational pull of, 255
  loss of interest in, 258
  lunar base, establishment of, 263, 265, 272–
    73
  lunar radio telescopes, 281–82
  making water on, 272
  mining on, for raw materials, 285
  moon-made solar-powered satellites, 288
  origin, theory of, 255
  outposts on, 282

  oxygen plant on, artist's conception of, **269**
  remaking, into useful products, 268, 272
  science accomplished on moon landings,
    258
  solar array networks on, 282, 284
  telescopes on, 273
  to test feasibility of manned mission to
    Mars, 272
  using raw materials of, to reduce costs, 281
Moon ship, living quarters in, **233**
Moravec, Hans, 309
Moritz, 59
Movies about space
  Frau Im Mond, 40, 41, **45**, **46**
  poster from *Frau Im Mond,* **40**
  2001: A Space Odyssey, 216, 257, 263
Multiple-combustion chambers, 57
Multistage rockets. *See* Rockets
Mutually Assured Destruction (MAD), 222

N

N-1 rocket, 95
National Academy of Sciences, xii, xiv, 79
National Aeronautics and Space
    Administration (NASA), 86
  budget problems, 96–97, 99, 128
  creation of, 86–88
  manned missions to Mars, long range plans
    for, 138
  space observatories, 17
National Aerospace Plane (NASP), 293
National Commission on Space, 140, 260, 263,
    272
National Oceanic and Atmospheric
    Administration (NOAA) vegetation
    index image, **245**, 249–50
National Space Development Agency
    (NASADA), 292
Nauchnoyie Obozrenie (Science Survey), 6, **6**
Navstar Global Positioning system satellite, **215**
Nebel, Rudolph, 43, 44, 46, **47**, 56
Neptune, 305
NERVA (Nuclear Engine for Rocket Vehicle
    Applications), 103, **105**
Newton, Sir Isaac, **16**, 303
Nimbus 7, 198
Noctis Labyrinthus, **166–67**
Noordung, Hermann, 218n
Nordhausen, 64, 67
North American Aerospace Defense
    Command (NORAD), 222–28
NOVA spacecraft, 91–92
Nuclear fission, 72
  See also Atomic bomb
Nuclear fusion, 288
Nuclear pulse engines, 306
Nuclear thermal rocket (NTR), 103
Nut, Egyptian Goddess of the sky, **14**

O

Obayashi Corporation, 169, 292
Oberth, Hermann, 19, **31**, 32–35, 43, 46, 47,
    47, 62, 202, 218n

advises Lang on movies, 40, 41
Die Rakete published, 33
letter to Goddard, 32
Observatories
Arecibo, 276
Gamma Ray, **275**
Lowell's, in Flagstaff, Ariz., 118
orbiting, 311
Solar Max, **264**
in space, 17
using the Moon as an astronomical
observatory, 273
Oil spill, Persian Gulf, **187**
Oil-well fires, Kuwait, **189**
Olympus Mons, **125**, 126, 169
O'Neill, Gerard, 286
Opel, Fritz von, 37, 39, 40
Opel Rak II, **38**
Opel Rak III, 40
Optical satellites, 224
"Orbital power," xiii
Orbital space stations, 234
Orbital transfer vehicle (OTV), 260, **261**, 263,
287
Ozone holes, **199**

P

Parsons, William, 276
Peenemünde, 52, 59, 103
assembly line, **58**
bombed by British, 64n
Photosynthesis, 296
Physical effects of microgravity, 142, 144, 145
Physics, 15
Phyton, 159
Pickering, William, 85, **85**, 86, 99
Pierce, John R., 234, 236, 250
Pioneer probes, 87, 97
Pioneer 10, 305
Planetary and Earth Sciences, 268
Plasma jets, 276
Plutonium, 72
Pobedas, 74–75
Polar Orbiting Earth Observation Missions
(POEM), 206
Polar-orbiting satellites, 247
Pollution, 190
Porter, Richard, 80
Probes
EOS, 204, 205, 206, 207
first Soviet, 93
Luna 15, 96
Mariner, 97, 122–28
to the moon's surface, 260
Pioneer, 10, 87, 97, 305
to search for extraterrestrial life, 305
Soviet Mars 2 & 3, 124
U. S., between 1969 and 1978, 97
Viking, 97, 126–32, 138, 299
Procaryotes, 184
"Progress," **174**
Project Amazon, 197
Propagules, 171
Proton, 290
Proxima Centauri, 305

Psychological problems of space crews, 148–
52
Ptolemeic model of the universe, 16
Ptolemeic Projection, **176**, 177
Pulsars, 276

Q

Quasars, 276, 311

R

Radiation belts, 85
Radio links, 57
Radio telescopes, 276, **276**, 281
to listen for extraterrestrial life, 304–305
lunar, 281–82
Rain forests, 199
acid rain, effects on, 191, 192
Amazon, 194–200, **195**
biological diversity of, 197
Brazilian government attempts to rescue,
198
ecosystem of, 197
effects of destruction of, on global climate,
200
Raketenflugplatz, 54
Raketenrummel, 54, 55
Ramanathan, Veerabhadran, 200, **201**, 202
RAND Corporation
studies on spy satellites, 219
Reagan, Ronald, 139, 140
Recoilless rocket launcher, 27
Red spruce needle, 194, **194**, **195**
Redstone, 78, 79, 86, 90
Relay communications satellite, **237**
Repulsor rockets, 47, 54–56
Revelle, Roger, 185
Rhyolite, 224
Ride, Sally, xii, 140
Ride Report, 204, 260, 263, 272
Riedel, Klaus, 43, 47, **47**, 54, 55, 70
Ritter, Franz, **47**
Robbie the Robot, 146
Robotics, 87, 97
research in, 132, 134
Rock, Barrett, 190, **190**, 192, 194
Rocket car, 37, **38**, 39
"Rocket into Planetary Space, The," 218n
Rockets
A-3, 60
A-4, 60
A-4 (V-2), 51
Aggregat-1 (A-1), 59
Atlas, 295
Centaur, 295
Chinese Long March, 200
early building and testing of, 25
first built at Peenemünde, 59–60, **61**, 62
first conceived by Tsiolkovsky, 7
first liquid fuel rocket launched, 41, **42**
GIRD X. GIRD, **68**, 69
Goddard's application for, **24**
liquid-fuel, 54
MIRAK, 54

Model B, **34**, 35
Model E, 35
in movie *Frau Im Mond,* 45
multistage, 44
N-1, 95
nuclear-electric propulsion, **102**
problems with early, 43–44
Redstone, 78, 79, 86, 90
Repulsor, 54, 55–56
Sander, 39
Saturn V, 95, 96, 97
Tsiolkovsky 's ideas on multistage rockets,
26
V-2, **50**, 60, 62
Vanguard, 79, 80, 82, **83**, 84
as weapons, 26, 27
Rocket ships, xiii
Romnenko, Yuri V., **142**
Roosevelt, Franklin, 72
Round the Moon, 6n
illustrations from, **8**, **9**
Russell, James, 116
Russia. *See* Soviet Union
Russian Academy of Sciences, 26

S

Sagan, Carl, 122, 165
and communicating with aliens, 308
Salyut and Mir space stations, 100, 260
Sander, Friedrich, 37, 39
Sänger space plane, 290, 292–93
Santy, Patricia, 144
Satellite cameras, 242
Satellite Communications Repeater (SCORE),
234
Satellite dishes, 239
Satellite maps
to track locust swarms, **250**
Satellite receiving stations, Niger, 247
Satellites, 75, 80, 212–51
Brilliant Pebbles, **221**, **223**
costs for building "sunsats", 286
DSP, **229**, **230**
ERS, **205**
"Ferret," 224
first American, *Explorer I,* 85, **85**
first communication, *Echo,* 87
first efforts at developing, 78–79
first weather, *TIROS I,* 87
geo-positioning, 249–50
LaCrosse, 224
Landsat, 126–28, 190, 191, 192, 247
meteorological, 198
to monitor rainfall, weather patterns and
crop growth, 246–49
moon-made solar-powered satellites, 288
Nimbus 7, 198
as orbital debris around Earth, 214
photos of Soviet SU-27, **220**
polar-orbiting, 247
Ryolite, 224
SIGNIT, 224
solar-powered (sunsats), 286
solar power station, 282, **282**, **283**
Sputnik, 37, 70, 80, **81**, 82, **212**

Satellites (*cont.*)
　types
　　communications, xiii, 213, 214, 216, 218–19, 236–41
　　first, SCORE, 234
　　first launched, 218
　　linked to cable television systems, 239, 241, 242
　　Telstar I, **236**, 237
　　navigational, 213, **215**
　　remote sensing, 213
　　spy, 213, 214, 219–22
　　　military relay, 224
　　weather, 87, 213, 214, 228, 242–47
　　spin-scan camera to watch global weather systems, 242, 244
Saturn rockets
　Saturn V rocket, 95, 96, 97
　super *Saturn*, **257**
Scablands of Washington State, 126–28, **127**
Schiaparelli, Giovanni, 117, 118, 121
　map of Mars, **117**
Schirra, Walter M., Jr., **87**
Schmitt, Harrison, **254**, 258
Schneikert, Frederick P., **51**, 52, 55
Schwarzkopf, Gen. Norman, 231
Schweikert, Rusty, 179
Science fiction, 8, 9, 106
　books, 121–22, 216
　　by Jules Verne, **6**, **8**, **9**, **10**, **11**
　magazines, **120**, 122, **213**, 216
　movies, 40, 146, 216, 257
Science Wonder Stories, **120**, 122
Scripps Institution of Oceanography, 201
Seamans, Robert, 88
Sea of Tranquillity, 265
Selenopolis, 299
Senegal, **246**, **247**
Sergeant missiles, 86
Setzer, Alberto, 198
Sex in space, 144–45
Shakespeare, William, 212
Sharp, Dennie, 239
Shepard, Alan B., Jr., **87**, 90
Sherskevsky, Alexander, 44, 46
Shimizu Corporation, 286, 292
　lunar base plan, **298**
　space hotel, **259**
Shockley, William, 234
Shuttle Remote Manipulator System, 264
Shuttle. *See* Space shuttles
SIGINT, 224
Skylab, 146, 150
Slash-and-burn expansion policy, Brazil, 196
Slave labor, rocket assembly plants, 67
Slayton, Donald K., **87**
Smith, Michael J., 99–100
Smithsonian Institution
　gives Goddard grant, 25
　publishes Goddard's paper on rocketry, 27
Smokestacks, electric power plant, **192**
Sociobiology, 196
Solar concentrator, **268**
Solar Max, **264**
Solar-powered satellites (SPS), 286
Solar power station, in space, 282, **282**, **283**
Solar sailing, **48–49**, 106, **107**, **108**, **109**

Solid-fuel rocket development, 86
Soviet SU-27 Flanker, **220**
Soviet Union
　explodes atom bomb, 74
　launches first liquid-fuel rocket, 69
　and launch of *Sputnik*, 80, 81, 82
　missile development in, 74, 75
　research on effects of microgravity, 142, 144, 145, 148–52
　Soyuz program, 91, 92, 93, 95
　space programs, 17
　space stations, 100, 142, **143**, 260
Space age
　birth of the, 44–47
　important changes from the, xii–xiii
Space Age, book and television series
　development of, xi
Space agencies, 17–19
　European, 18, 206
　INPE, 198
Space Devastator, **213**
Space exploration
　in Germany, 35, 36, 37
　U. S. master plan for, 90
Space Exploration Initiative, 272
Space factories, 286
Spaceflight, 260
Space Industries International, 297
Space junk, 214
Space planes, 290, 292–95
Space programs, 15–19, 100
Space science, xii
Space service platform, **266–67**
Space ships
　first docking of, in space, 91
　to Mars, **155**, **160**, **161**
　Mars Observer , 163
　unpiloted, **289**
Space shuttles, **98**, 99, **101**, 290, 292
　Challenger, 99–100, **100**
　Columbia, 97, 228
　costs of, 288, 290
　docking at space station, **262**
Space societies
　in USSR, 37
　Verein für Raumschiffarht (VfR), 36, 37, 39
Space stations, 204
　designs for, 146–48
　Earth-orbiting, 260, 263
　Freedom, 111, 144, 204
　plans for, 140
　Soviet, 100, 142, **143**, 260
　Wernher von Braun's concept of, **147**
Space Studies Institute, 286
Space suits, 145–46
Space Task Group, 138
Space Transportation System (STS).
　See Space shuttle
Space walk, first, 91
Speer, Albert, 63
Spielberg, Steven, 297
Spin camera weather satellite
　NOAA vegetation image, **245**
　view of global weather systems, 243, 244
Sputnik, 37, 70, 234
　first launch of *Sputnik I,* 80, **81**, 82, 212
Staehle, Rob, 106–7

Stalin, Joseph, 290
Stapledon, Olaf, 299
Stars
　47 Tucanae, **272**
　Orion Nebula, **272**
"Star Wars" research, 103
Steam engine, invention of, 178
Stone, Edward, xii
Straits of Gibralter, **180–81**
Strassman, Fritz, 72
Stromatolites, 137
Sudety Mountains, Poland, **193**
Sullivan, Walter, 80, 81
"Sunsats," 286
Suomi, Werner, **242**, 244, 250
Symbiogenesis, 183, **184–85**
System Z, 204

T

Taurus Littrow, 258, 265
TDRS system, **225**, **226–27**
"Tele-family of man," 240
Telescopes
　and communication with satellites, 228
　Hubble Space Telescope, **272**, 273, **274**, 276, 281, 311
　infrared, 281
　lunar, **270–71**
　radio, 276, **276**, **281–82**
Telstar I, 236, **236**, 237
Temperature
　average, surface of Earth, **203**
Termite microorganisms, **184**
"Terraforming," of Mars, 309–10
Thematic Mapper, 192
Thiel, Walter, 61
Thomas, Lewis, 302, 308, 309
TIROS I, 87
Titanium, 265
Tomahawk cruise missile, **110**
Transistors, 232, **234**
Traveling in space
　costs of, 288, 290
Tropical storm SAM
　orbital photo of, **241**
Tsander, Fridrikh, 37, 290
Tsiolkovsky, Konstantin, **5**, **12**, 19, 69, 103, 106, 159, 202, 218n, 299, 303, 309
　drawing of "hothouse" in space, **157**
　first calculations for ascent into space, 5–6
　and harnessing of solar energy, 284
　ideas on multistage rockets, 26
　images of alien life-forms, 307
　letter to a friend, **12**
Tukhachevesky, Marshall, 70
Tupolev, Andrei, 69, 70
Twenty Thousand Leagues Under the Sea, 6n
2001: A Space Odyssey, 216, 257, 263

U

"Ultimate Migration, The," 20, 106, 309
　envelope for, **21**

U. S. Army
  finds V-2 technology at Peenemünde, 64

V

V-2 rockets, **50**, 51, 51n, 52–55
  early problems with, 56
  remains of, detonated in London, **66**
  tested at White Sands, 76
  test launch of, **61**, 62
Valier, Max, 36, 37, **38**, 40, 54
Valles Marineris, **125**, 126, 138, 163, **166–67**, 168, 169
Van Allen, James, 78, 85, 99
Van Allen Radiation Belts, 85
Vanguard, 79, 80, 82
  explodes during launch, **83**, 84
Venera I, 93
Vengeance Weapon-1, 51n
Venus, **i**, **ii–iii**, **iv–v**, **vi–vii**, 196
Verein für Raumschiffarht (Rocket Society), 36, 37, 39, **47**, 54
Vernadsky, Vladimir, 159, 179, 182, **182**, 183
  anticipates Space Age, 182
Verne, Jules, 6, 33
Very Large Array, 276, **276**, 281

Vespucci, Amerigo, 177
Viking probes, 97, 126–32, 138, 299
  channels on Mars, **127**
Voice cable, 232
Voskhod missions, 91, 92
Vostok, 142
Voyager, 97, 305

W

Walcott, Charles D., 25
Walker, Joe, 293
Wallace, Alfred Russel, 194
Walter, William "Chip," xii
Warner, Charles Dudley, 242
War of the Worlds, The, 22
Watt, James, 178
Weapons
  development, by Germany , 51–67
  recoilless rocket launcher, 27
Weather balloons, 234, 236
Webster, Arthur G., 27
Wells, H. G., 22, 161, 174, 281, 299, 307
Whirlpool Galaxy M51, 276, **277**
White Sands Missile Range, 75–76, 78

Wildlife
  destruction of, by acid-rain, 192
Wiley, Carl, 106
Wiley, Lynn, 145
Wilson, E. O., 196, 197
Winged bean, 197
Wireless World, 216, 224, 232
World Space Foundation, 107
WQED, Pittsburgh, xi, xii, xiv
Wright Brothers, 5, 6

X

X-15 rocket-powered aircraft, 293, **294**
X-30, 293
X-ray universe, 311

Y

Yeager, Chuck, 293

Z

Zero gravity. *See* Microgravity
Zoroaster and the Magi, **14**

The Author

WILLIAM J. (CHIP) WALTER is a documentary filmmaker, screenwriter, and former network bureau chief for CNN. He recently moved from Los Angeles and currently lives with his wife, Mary, and young daughter, Molly, in Pittsburgh, where he is National Programming Executive with WQED Television.